ELECTRON SPIN RESONANCE IN FOOD SCIENCE

ELECTRON SPIN RESONANCE IN FOOD SCIENCE

Edited by

ASHUTOSH KUMAR SHUKLA

Ewing Christian College, Allahabad, Uttar Pradesh, India

ACADEMIC PRESS

An imprint of Elsevier

elsevier.com

Academic Press is an imprint of Elsevier
125 London Wall, London EC2Y 5AS, United Kingdom
525 B Street, Suite 1800, San Diego, CA 92101-4495, United States
50 Hampshire Street, 5th Floor, Cambridge, MA 02139, United States
The Boulevard, Langford Lane, Kidlington, Oxford OX5 1GB, United Kingdom

Notices
Knowledge and best practice in this field are constantly changing. As new research and
experience broaden our understanding, changes in research methods, professional practices,
or medical treatment may become necessary.

Practitioners and researchers must always rely on their own experience and knowledge
in evaluating and using any information, methods, compounds, or experiments described
herein. In using such information or methods they should be mindful of their own safety
and the safety of others, including parties for whom they have a professional responsibility.

To the fullest extent of the law, neither the Publisher nor the authors, contributors, or
editors, assume any liability for any injury and/or damage to persons or property as a
matter of products liability, negligence or otherwise, or from any use or operation of any
methods, products, instructions, or ideas contained in the material herein.

British Library Cataloguing-in-Publication Data
A catalogue record for this book is available from the British Library

Library of Congress Cataloging-in-Publication Data
A catalog record for this book is available from the Library of Congress

ISBN: 978-0-12-805428-4

For Information on all Academic Press publications
visit our website at https://www.elsevier.com

Working together
to grow libraries in
developing countries

www.elsevier.com • www.bookaid.org

Publisher: Nikki Levy
Acquisition Editor: Nina Bandeira
Editorial Project Manager: Mariana Kuhl
Production Project Manager: Nicky Carter
Designer: Maria Ines Cruz

Typeset by MPS Limited, Chennai, India

DEDICATION

Dedicated to my grand parents.

CONTENTS

LIST OF CONTRIBUTORS

K. Akram
University of Sargodha, Sargodha, Pakistan

V. Bercu
University of Bucharest, Magurele (Ilfov), Romania

M. Cutrubinis
Horia Hulubei National Institute of Physics and Nuclear Engineering (IFIN-HH), Măgurele, Romania

O.G. Duliu
University of Bucharest, Magurele (Ilfov), Romania

U. Farooq
University of Sargodha, Sargodha, Pakistan

G.P. Guzik
Institute of Nuclear Chemistry and Technology, Warsaw, Poland

S. Iravani
Isfahan University of Medical Sciences, Isfahan, Iran

K.P. Mishra
Ex Bhabha Atomic Research Center, Mumbai, Maharashtra, India

K. Nakagawa
Hirosaki University, Hirosaki, Japan

C.D. Negut
Horia Hulubei National Institute of Physics and Nuclear Engineering (IFIN-HH), Măgurele, Romania

A. Shafi
University of Sargodha, Sargodha, Pakistan

A.K. Shukla
Ewing Christian College, Allahabad, Uttar Pradesh, India

A.I. Smirnov
North Carolina State University, Raleigh, NC, United States

PREFACE

This book intends to describe the applications of electron spin resonance (ESR) spectroscopy in the area of Food Science. Electron paramagnetic resonance (EPR) and ESR: both names are equally valid for the technique. A relatively new term, electron magnetic resonance (EMR), is also gaining popularity. Authors have freely adopted the title and style of their chapters in this book. These expert authors are from different disciplines, and accordingly interdisciplinary character has come up in a natural way. The emphasis is on the application side, and mathematical details have been kept to a minimum.

This book contains eight chapters. Constantin Daniel Negut and Mihalis Cutrubinis have described standard ESR methods for detection of irradiated food in Chapter 1, Electron Spin Resonance Standard Methods for Detection of Irradiated Food. In Chapter 2, Electron Paramagnetic Resonance Investigation of the Free Radicals in Irradiated Foods, Octavian G. Duliu and Vasile Bercu have described quantitative ESR analysis of the time and temperature dependence of free radicals in irradiated foods. In Chapter 3, Electron Spin Resonance Techniques in the Quality Determination of Irradiated Foods, Kaushala Prasad Mishra has discussed quality issues related with food irradiation, and emphasized the role of ESR in quality determination. Chapter 4, Electron Spin Resonance Detection of Irradiated Food Materials is an attempt by Grzegorz Piotr Guzik and myself to introduce the actual subject matter of this book, with some examples from a wide range of food materials. Kashif Akram, Umar Farooq, and Afshan Shafi have especially covered ESR identification of irradiated fruits and vegetables in Chapter 5, Electron Spin Resonance Spectroscopy for the Identification of Irradiated Fruits and Vegetables. Alex I. Smirnov has described ESR applications to beverages in Chapter 6, Electron Paramagnetic Resonance Spectroscopy to Study Liquid Food and Beverages. Siavash Iravani has reviewed the ESR of irradiated drugs and excipients for drug control and safety in Chapter 7, Electron Spin Resonance of Irradiated Drugs and Excipients for Drug Control and Safety. In Chapter 8, Free Radicals in Nonirradiated and Irradiated Foods Investigated by Electron Spin Resonance and 9 GHz Electron Spin Resonance Imaging, Kouichi Nakagawa has covered ESR imaging as it is applied to the investigation of food items. I learnt many things from these chapters, and hope that readers will also enjoy reading it in a fruitful way.

I sincerely thank to Nina Bandeira, Food Science Acquisitions Editor, Elsevier, for giving me an opportunity to present this book to readers. I wish to thank Mariana Kühl Leme, Editorial Project Manager, Elsevier, for providing all the support during the development of this project. I thank the authors for taking time out of their busy academic schedules to contribute to this book. It is their cooperation which led me to develop this project. My special thanks to anonymous reviewers for their contributions to improve the quality.

I am grateful to Prof. Ram Kripal, Head, Department of Physics, University of Allahabad, who introduced me to ESR spectroscopy. My sincere thanks to Prof. Raja Ram Yadav, Department of Physics, University of Allahabad, Dr. M. Massey, Principal, Ewing Christian College, Allahabad, and my colleagues for their constant encouraging remarks and suggestions during the development of this book project.

It is difficult to express my gratitude in words to my parents who have blessed me to complete this task. My brother Dr. Arun K. Shukla, Department of Biological Sciences and Bioengineering, Indian Institute of Technology, Kanpur has always been there to help me. My special thanks are also due to my wife Dr. Neelam Shukla, my daughter Nidhi, and son Animesh for their patience during this work.

Ashutosh Kumar Shukla

Allahabad, India

December 2016

CHAPTER 1

ESR Standard Methods for Detection of Irradiated Food

C.D. Negut and M. Cutrubinis
Horia Hulubei National Institute of Physics and Nuclear Engineering (IFIN-HH), Măgurele, Romania

Contents

1.1 INTRODUCTION

Although the biocide effect of ionizing radiation has been known since the late 19th century, the first food irradiation facility was commissioned only in 1959. In 1980, the Joint FAO/IAEA/WHO Expert Committee on the wholesomeness of irradiated food concluded that irradiation of food up to an average dose of 10 kGy presents no toxicological hazard and introduces no special nutritional or microbiological problems [1]. Nowadays, irradiation is a regulated technology for preventing losses or for microbial decontamination of food. In 2011, the International Organization for Standardization (ISO) issued a standard for treatment of food by ionizing radiation [2]. According to the Irradiated Food Authorization Database [3], irradiation of food has been approved in more than 50 countries. Regulations regarding categories of food approved for

---

Electron Spin Resonance in Food Science.

1

irradiation and maximum permitted doses vary widely from country to country. In many countries of the European Union, only dry vegetables and spices can be treated by ionizing radiation at a maximum dose of 10 kGy, while in Brazil any category of food can be irradiated at any dose.

Regulations across the world require accurate labeling in order to inform the consumers whether foodstuffs or ingredients within them have been irradiated. Thus, methods were developed for the detection of irradiated food which can be used in two ways: to see if irradiated foodstuffs are correctly labeled, or if foodstuffs labeled as irradiated are really irradiated. Most of the research on the detection of irradiated food was done between 1985 and 1995, resulting in the adoption of 10 standards by the European Committee for Standardization (CEN). The standards are complementary techniques, allowing the detection of irradiation treatment for a wide variety of foods. Among these, three are based on ESR spectroscopy.

Electron spin resonance (ESR) detects free radicals induced by ionizing radiation and trapped in the dry parts of irradiated food. Their stability is related to the moisture content and the crystallinity of the solid region of food. In principle, ESR signals (as well as any other parameters used in the detection of irradiated food) induced by ionizing radiation should be absent in nonirradiated food and specific to irradiation, meaning that it cannot be induced by any other food processing methods. It should appear at the usual doses applied to the specific food under test, and should be stable at least over the storage life of the irradiated food [4]. Ideally, the signal intensity monotonously increases with the dose, making an estimation of the applied dose possible. This is the case of very stable free radicals such as carbonate radicals in the hydroxyapatite crystal of bone. According to the practical criterion that a detection method should apply to a wide range of food types, there are standardized ESR-based detection methods for food containing bone [5], cellulose [6], and crystalline sugar [7].

1.2 STANDARDIZED ESR DETECTION METHODS

1.2.1 Basic Instrumentation

Food irradiation control by ESR is usually performed in X-band (about 9.1–9.7 GHz) by continuous wave (CW) mode. The main task is to find radiation-induced signals with characteristic shapes and g factors. The value of the g factor is obtained from the values of the resonant magnetic field and microwave frequency. The microwave frequency can be measured with high precision by a frequency counter. The magnetic field is usually determined by means of a Hall sensor. A higher accuracy can be achieved by using a nuclear magnetic resonance (NMR) gaussmeter. Another way to determine the g factor is to use a reference material with a well-known g value. The reference is measured together with the sample such that their spectra are obtained for the same frequency and magnetic field [8]. The unknown g factor is derived from the resonant magnetic field and the g factor of reference. Ideally, the g factor of the reference differs from that of the sample enough that their spectra will not overlap. A powder of Mn^{2+} ions in CaO is a very convenient reference material. Its ESR spectrum consists of a hyperfine sextet with a g factor of 2.0010. The lines are spaced by $8-9\,mT$, thus both the third (g about 1.9810) and the fourth lines (g about 2.0330) can be used as a reference. In general, ESR signals found in irradiated food are large. In the case of sugars, the spectrum width can reach $10\,mT$, thus in acquiring the spectrum a sweep of $20\,mT$ around the central field ($342\,mT$ for $9.5\,GHz$) is recommended.

One of the advantages of the ESR method is the simplicity of sample preparation. Usually, a scalpel is enough to obtain dry parts of the food by removing soft tissues which can affect the signal. When seeds from fresh fruit are being analyzed, additional common instruments and materials are needed to separate the seeds from the pulp, such as an electric blender and purified water to dilute the pulp and filter paper to absorb residual water from the seeds. The high moisture content of samples can cause difficulties in tuning the spectrometer resonator. In such cases it is recommended to dry the samples using a freeze dryer or a laboratory vacuum oven at a temperature up to 40°C. Excessive heating may drastically reduce the radiation-induced signal and, at the same time, induce free radicals that are not specific to irradiation.

1.2.2 Food Containing Bone

The first ESR study on radiation-induced free radicals on bone was reported in 1955 [9]. Since then, many papers have been devoted to explaining their structure and investigating their stability. Hydroxyapatite, $Ca_{10}(PO_4)_6(OH)_2$, is the main mineral component of bone. As it has a complex structure, it can accommodate different free radicals upon irradiation, but the most common are those of carbonate and oxygen. Among these, carbonate radical CO_2^- is the most stable and its ESR signal dominates the spectrum. It has fairly good radiation sensitivity so that its signal can be used in dosimetry, dating of fossils, and detection of irradiated food containing bone [10].

Nonirradiated bone can exhibit a low intensity, broad symmetrical single line with a g factor (g_c) of about 2.005 attributed to the marrow [11]. Thus, it is recommended that marrow is removed when possible. Sometimes, the ESR spectrum is complicated by the presence of manganese (Mn^{2+}) producing a sextet with a hyperfine splitting of about 8.5 mT, typical for manganese as an impurity in carbonates [12]. Fig. 1.1A shows the spectrum of nonirradiated hake bone where the broad singlet overlaps the third line of Mn^{2+}, while the fourth line (marked by a dashed rectangle) is well resolved.

Irradiated bone shows a narrow, apparently axial asymmetric line with a $g_\perp$ value of 2.002, and a $g_{||}$ value of 1.998, which easily distinguishes it from the native one. Its intensity and stability depend on the crystallinity of bone [13]. In general, detection is possible at doses higher than 0.5 kGy, which covers the typical dose range for the irradiation of meat and fish: 0.5–8 kGy. Fig. 1.1B shows the spectrum of 2 kGy gamma irradiated hake. The irradiation-induced signal is well developed and clearly distinguished from the native one. Even if the fish bone is of low mineralization, the irradiation signal will be stable for more than 2 years (Fig. 1.1C).

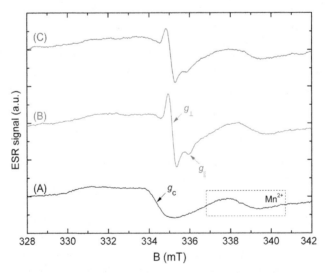

Figure 1.1 ESR spectra for nonirradiated (A), and 2 kGy gamma irradiated (B) hake bone recorded immediately after irradiation (same gain, 0.4 mT modulation amplitude and 5 mW microwave power). The irradiation-induced signal is well developed and clearly distinguished from the native one, even 2 years after irradiation (C).

The method was validated by inter-laboratory tests on beef bones, chicken bones, and trout bones at doses of 2 kGy and higher. Other food-stuffs for which the method has proved successful are frog leg [14], egg (shell) [15], and shellfish [16,17]. An evaluation of the irradiation dose is possible by using an additive dose method and a carefully chosen dose – response curve [18].

1.2.3 Food Containing Cellulose

Cellulose is a major component of plant cell walls and it is present in virtually all food of plant origin. Some parts of plants, such as fruit pulp, have a high moisture content which increases the mobility of radiation-induced radicals so that their depletion is very rapid. Other parts of fruit (such as seeds or shells) with a low water content can be used for detection of radiation treatment. The first ESR study on radiation-induced radicals in cellulose dates back to the 1960s [19], but the process of radical formation in gamma-irradiated cellulose has been convincingly explained much later [20]. Ionizing radiation can induce at least two specific radicals in pure cellulose, resulting in a spectrum consisting of two signals from an unpaired electron interacting with one proton (a doublet with hyperfine splitting a_H of about 2.6 mT) and with two protons (a triplet with a_{2H} of about 3.0 mT and intensity ratio of 1:2:1) [21,22].

Nonirradiated food containing cellulose presents a single, slightly asymmetrical line, with an apparent g_c of about 2.004 (Fig. 1.2A) which is attributed to the semiquinone radicals [23]. Its intensity usually increases greatly with the absorbed dose. After irradiation only the triplet appears, with the satellite lines separated by approximately 6.0 mT. They are generally very weak compared with the central, nonspecific-to-irradiation signal which overlaps the central line of the triplet, thus they can be missed if an improper gain is used. Fig. 1.2B shows the spectrum of laurel leaves irradiated at 5 kGy. An irradiation specific signal is hardly visible if the gain is chosen such that the central signal is recorded full scale. The inset of Fig. 1.2 shows the same signal recorded out-of-scale where the pair of satellite lines, separated by 6.0 mT, are clearly distinguished from the background. The specific signal saturates at low microwave power, thus it is recommended to work at powers lower than 1 mW.

The presence of Mn^{2+} paramagnetic ions can seriously limit the detection of irradiated food containing cellulose. Although the spacing between the third and the fourth lines of the manganese sextet is about 8.5 mT, sometimes these lines are strong enough such that only the left specific

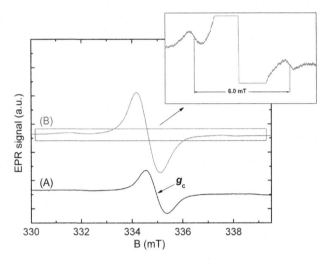

Figure 1.2 ESR spectra for nonirradiated (A), and 5 kGy gamma irradiated (B) laurel leaves (same gain, 0.5 mT modulation amplitude and 0.8 mW microwave power). The inset shows the spectrum (B) recorded at a 10 times higher gain in order to enhance the specific-to-irradiation signal.

line of the cellulose radical can be detected, at 3 mT from the central line [11]. In Fig. 1.3 the spectrum of pumpkin seed husk irradiated at 5 kGy is shown; only the left satellite line (marked by a red circle) of the cellulose radical is discernible, but it can be used as proof of irradiation.

The "cellulose" radical can be detected in different parts of plants [24–29] and mushrooms [30], spices [31,32], sauces [33], and even in packaging material [34]. The intensity of the specific signal is dependent on the species and the observed part of the food being analyzed. Its stability is strongly influenced by the crystallinity of cellulose and the moisture content of the sample, thus the absence of a specific signal is not proof that the sample has not been irradiated [35]. The minimum dose at which irradiation can be detected varies widely (from hundreds of Gy to a few kGy). The method was validated by inter-laboratory tests for pistachio nuts (for doses of 2 kGy and higher), paprika powder (5 kGy), and fresh strawberries (1.5 kGy).

1.2.4 Food Containing Crystalline Sugar

Different types of sugar (fructose, glucose, sucrose, etc.) present in food can give rise to specific ESR signals upon irradiation. Spectra are large (overall spectrum width varying from 7 to 10 mT) and very complex as a result

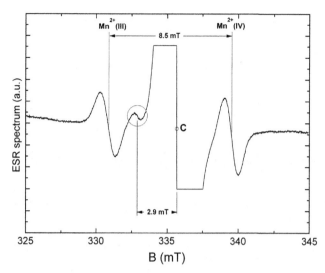

Figure 1.3 The ESR spectrum of pumpkin seed husk registered 2 months after gamma irradiation at 5 kGy (modulation amplitude 0.5 mT and 0.8 mW microwave power). The right line of the cellulose radical is overlapped by the fourth line of Mn^{2+}, while the left one (marked with a red (gray in print versions) circle), separated by about 3 mT from the central nonspecific signal (C), is well resolved and can be used as proof of irradiation.

of overlapping lines with different intensities and linewidths. Since the beginning of the 1960s [36] there have been many attempts to elucidate the structure of sugar radicals and explain their contribution to the ESR spectrum. Many studies were focused on irradiated sucrose because some of the radicals have good stability at room temperature, making table sugar a potential emergency dosimeter. Sugar radicals are also of interest for biochemistry. Some of the defects induced by ionizing radiation in DNA are related to specific sugar radicals, thus a good understanding of simpler radicals such as those of sucrose may provide insight into damage induced in the DNA by radical reactions [37]. Thus, it is not surprising that sucrose is one of the most studied carbohydrates. Despite all the efforts over the past 50 years, sugar radicals are not completely understood. However, recent studies agree that at room temperature the spectrum is dominated by the signals from three stable radicals [38,39]. Experimental and calculated data suggest these are carbon-centered radicals in hyperfine interaction with protons [40].

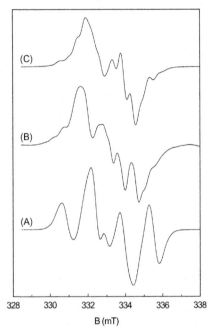

Figure 1.4 The ESR spectra of standard (analytical purity) fructose (A), glucose (B), and sucrose (C) gamma irradiated at 5 kGy (same gain, 0.2 mT modulation amplitude and 1 mW microwave power). "Sucrose-like" spectra are the most common in the dry parts of irradiated fruits.

Fig. 1.4 shows the spectra of fructose, glucose, and sucrose (of analytical purity) gamma irradiated at 5 kGy. Similar spectral shapes may appear in irradiated food containing crystalline sugars, but the most common is that of sucrose which can be regarded as a combination of fructose and glucose [41].

The composition, crystallinity, and quantity of sugars in dried fruit vary according to species and which parts are analyzed; thus, variations in the shape and intensity of the ESR radiation–induced signal can be observed even in the same sample. Moreover, there are reported variations in the spectral shape of irradiated standard (laboratory grade) sucrose for different ESR spectrometers [42]. Nevertheless, the appearance of a multicomponent ESR spectrum with similar features to those of sugar radicals indicates radiation treatment. Two different ESR signals may be observed in the dried parts of nonirradiated fruits containing crystalline sugars: a central, broad singlet with a g_c of about 2.004, and the sextet of Mn^{2+}. Upon irradiation, the characteristic signal of sugar radicals dominates the

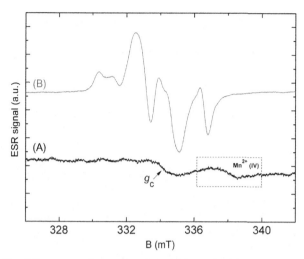

Figure 1.5 The ESR spectra for nonirradiated (A), and 2 kGy gamma irradiated (B), dried figs recorded at 2 months after irradiation (0.4 mT modulation amplitude and 5 mW microwave power; the nonirradiated sample was recorded at a four times higher gain than the irradiated one). The "fructose-like" signal (B) dominating the spectrum immediately proves irradiation; in normal storage conditions it can last more than 1 year.

spectrum even at low doses, allowing easy identification of radiation treatment. Because some lines are narrow, it is recommended to use modulation amplitudes lower than 0.4 mT, in order not to miss some features of the characteristic spectrum. Fig. 1.5 shows the spectra of nonirradiated and 2 kGy gamma-irradiated dried figs. The nonirradiated sample exhibits a broad native singlet (g_c of about 2.004) overlapping the third line of Mn^{2+} evidenced by the fourth one, marked by a dashed rectangle in Fig. 1.5A. The irradiated sample shows a very strong, broad (about 9.5 mT) signal dominated by a "fructose-like" spectrum (see Fig. 1.4A) that immediately proves irradiation. Storage in normal conditions of temperature and humidity will not affect the signal for more than 1 year. Detection is primarily limited by the amount and crystallinity of sugar in the sample. Hydration and heating can fade the radiation-induced signal, thus its absence does not constitute evidence that the sample is nonirradiated. When needed, radiation sensitivity (related to sufficient crystalline sugar, low moisture) can be tested by irradiating portions of the sample at different doses in order to evaluate the lower detection limit.

The method was validated by inter-laboratory tests for dried fruit: papaya, raisin, mango, and fig for doses of 0.5 kGy or higher. "Sugar-like" signals were reported not only for dried fruits, but also for some species of mushrooms [30,43,44], soybean-based powdery sauces [45], medicinal herbs [46,47], fruit syrups [48], and grains [49].

1.3 IMPROVEMENTS TO ESR DETECTION METHODS

The low crystallinity of carbohydrates (cellulose and sugars) strongly decreases the stability of radiation-induced radicals in foods of plant and mushroom origin, resulting in very weak or absent specific-to-irradiation ESR signals. For many kinds of spices and dried herbs only a strong increase in the intensity of the native, broad central singlet is observed upon irradiation [50,51]. Moreover, radiation induces no significant changes in its spectral parameters (g factor, linewidth, and line shape) [52], thus it cannot be directly used for irradiation detection. In such cases the three European standards cannot be applied. Over the last two decades, different improvements to the ESR techniques have been proposed in order to overcome this problem.

1.3.1 Thermal Annealing

A nonspecific radiation-induced signal is very unstable in comparison with a nonirradiated one, and this peculiarity can be used to provide evidence of radiation treatment. In Fig. 1.6 the kinetics of central signal intensity (considered as peak-to-peak amplitude) are shown, recorded at room temperature over 45 days, for nonirradiated and 5 kGy gamma-irradiated dried green tea leaves. The irradiated sample shows a strong decay which can be described by an exponential decay with a half time of about 2 days. This behavior clearly distinguishes it from the nonirradiated sample that has a rather constant intensity, even with a small increase in the first 10 days. Such fast decay of the central signal indicates irradiation. However, after 20 days (corresponding to 10 half times) the signal decay of the irradiated sample is too slow to differentiate it from the nonirradiated one. For routine control, observing the decay over several days is not a practical method.

In order to shorten the analysis time, thermal annealing can be used to evaluate the stability of the nonspecific signal [53]. In many cases, isothermal annealing at about 60°C for a few hours will show a strong decrease in the intensity of the central signal for irradiated samples. However, a strong

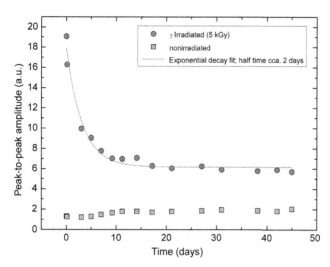

Figure 1.6 The variation in central ESR signal intensity (peak-to-peak amplitude) for nonirradiated and 5 kGy gamma-irradiated green tea leaves. Spectra were recorded at room temperature over 45 days. The irradiated sample shows a strong decay which can be described by an exponential with a half time of about 2 days. This behavior clearly distinguishes it from the nonirradiated sample in the first 20 days (about 10 half times) after irradiation. After 20 days the signal decay of the irradiated sample is too slow to be differentiated from the nonirradiated one.

decrease in the intensity of the signal upon heating cannot be used as proof of irradiation. Radiation treatment must be confirmed by other more expensive and time consuming techniques such as thermoluminescence (TL) [54]. Still, the method can be applied as a screening test taking into account that it is simple and fast. It is limited by the stability of the non-specific radiation-induced signal; if its half time is too short, applying the method after a period of time corresponding to several half times will not reveal any significant decrease in its intensity.

1.3.2 Saturation Behavior

Another proposed method is based on the fact that for certain dried plants the microwave power saturation behavior of the nonspecific central signal differs between nonirradiated and irradiated samples [55]. Fig. 1.7 shows an example of such behavior for nonirradiated and 5 kGy gamma-irradiated desiccated oyster mushroom. The irradiated sample shows a maximum around 9 mW, while the nonirradiated one follows an

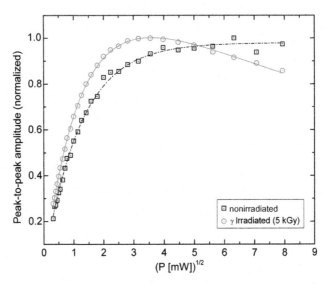

Figure 1.7 Saturation behavior of the central ESR signal for desiccated oyster mushroom; the irradiated sample shows a maximum around 9 mW, while the nonirradiated one follows an exponential saturation curve.

exponential saturation curve. However, other studies suggest this peculiarity is species dependent and cannot be generalized [49,56].

1.3.3 Spin Probe

Nitroxides are stable radicals with a high sensitivity to their micro environment which strongly influences their ESR spectra. They are used as spin probes to study diamagnetic systems with many applications in biology and material science. Because irradiation affects the structure of the plasma membrane, the spin probe method can be used when radiation-induced radicals have disappeared. The method was successfully tested for detection of 2.5 kGy gamma-irradiated seeds 8 months after irradiation [57]. The technique is complicated [58] and requires highly skilled operators, thus its potential as a routine method is low.

1.3.4 Alcoholic Extraction

Detection sensitivity can be improved for the soft tissue of fruit by alcoholic extraction to remove water and other constituents that can affect the specific ESR signal. The insoluble solid fraction is extracted from the fruit pulp and measured after drying at room temperature. By this method, irradiation can be detected at doses as low as 100 Gy; a fair estimation of the applied dose can be obtained by the additive dose method [59]. In the last 2 years, the method was successfully tested not only on fresh fruits, but also on fruit juice [48], fresh vegetables [60], dried spices [61], and sauces [33]. All these studies point out that alcoholic extraction gives much better results than classical pre-treatment methods such as freeze-drying. Sample preparation is not complicated, and detection is performed following European standards for food containing cellulose and crystalline sugar, therefore this method has great potential to be standardized.

REFERENCES

[1] R.A. Molins (Ed.), Food Irradiation: Principles and Applications, Wiley-Interscience, New York, 2001 (Chapter 1).
[2] ISO 14470, Food Irradiation – Requirements for the Development, Validation and Routine Control of the Process of Irradiation Using Ionizing Radiation for the Treatment of Food, 2011.
[3] Irradiated Food Authorization Database, <https://nucleus.iaea.org/ifa>, 2016 (accessed 24.03.16).
[4] E.M. Stewart, Detection methods for irradiated foods, in: R.A. Molins (Ed.), Food Irradiation: Principles and Applications, Wiley-Interscience, New York, 2001.
[5] EN 1786, Detection of Irradiated Food Containing Bone – Method by ESR Spectroscopy, 1996.
[6] EN 1787, Detection of Irradiated food Containing Cellulose by ESR Spectroscopy, 2000.
[7] EN 13708, Detection of Irradiated Food Containing Crystalline Sugar by ESR Spectroscopy, 2001.
[8] P. Höfer, Basic experimental methods in continuous wave electron paramagnetic resonance, in: M. Brustolon, E. Giamello (Eds.), Electron Paramagnetic Resonance: A Practitioners Toolkit, John Wiley & Sons, New Jersey, 2009.
[9] W. Gordy, W.B. Aard, H. Shields, Microwave spectroscopy of biological substances, I: paramagnetic resonance in X-irradiated amino acids and proteins, Proc. Natl. Acad. Sci. USA 41 (1955) 983–996.
[10] F. Callens, G. Vanhaelewyn, P. Matthys, E. Boesman, EPR of carbonate derived radicals: Applications in dosimetry, dating and detection of irradiated food, Appl. Magn. Reson. 14 (2) (1998) 235–254.
[11] J. Raffi, P. Stocker, Electron paramagnetic resonance detection of irradiated foodstuffs, Appl. Magn. Reson. 10 (1) (1996) 357–373.
[12] V. Bercu, C.D. Negut, O.G. Duliu, Detection of irradiated frog (*Limnonectes macrodon*) leg bones by multifrequency EPR spectroscopy, Food. Chem. 135 (4) (2012) 2313–2319.
[13] J.F. Diehl, Safety of Irradiated Food, second ed., CRC Press (1999), (Chapter 5).
[14] J. Raffi, J.C. Evans, J.-P. Agnel, C.C. Rowlands, G. Lesgards, ESR analysis of irradiated frogs' legs and fishes, Int. J. Rad. Appl. Instrum. A 40 (10) (1989) 1215–1218.
[15] B. Engin, H. Demirtaş, M. Eken, Temperature effects on egg shells investigated by XRD, IR and ESR techniques, Radiat. Phys. Chem. 75 (2) (2006) 268–277.
[16] J. Raffi, C. Hasbany, G. Lesgards, D. Ochin, ESR detection of irradiated seashells, Appl. Radiat. Isot. 47 (11–12) (1996) 1633–1636.

[17] E.M. Stewart, The application of ESR spectroscopy for the identification of irradiated Crustacea, Appl. Magn. Reson. 10 (1) (1996) 375–393.
[18] F. Bordi, P. Fattibene, S. Onori, M. Pantaloni, ESR dose assessment in irradiated chicken legs, Radiat. Phys. Chem. 43 (5) (1994) 487–491.
[19] J. Arthur, T. Mares, O. Hinojosa, ESR spectra of gamma-irradiated cotton cellulose I and II, Text. Res. J. 36 (7) (1966) 630–635.
[20] B.G. Ershov, O.V. Isakova, Formation and thermal transformations of free radicals in gamma radiation of cellulose, Bull. Acad. Sci. USSR 33 (6) (1984) 1171–1175.
[21] K. Sultanov, A.S. Turaev, Mechanism of the radiolytic transformations of cellulose 1. Formation and transformation of radicals, Chem. Nat. Compd. 32 (4) (1996) 570–576.
[22] M. Wencka, K. Wichlacz, H. Kasprzyk, S. Lijewski, S. Hoffmann, Free radicals and their electron spin relaxation in cellobiose. X-band and W-band ESR and electron spin echo studies, Cellulose 14 (2006) 183–194.
[23] B.A. Goodman, D.B. McPhail, D.M.L. Duthie, Electron spin resonance spectroscopy of some irradiated foodstuffs, J. Sci. Food Agric. 47 (1) (1989) 101–111.
[24] J.J. Raffi, J.-P.L. Agnel, L.A. Buscarlet, C.C. Martin, Electron spin resonance identification of irradiated strawberries, J. Chem. Soc. Farad. T. 1 (84) (1988) 3359–3362.
[25] M.A. Ghelawi, J.S. Moore, N.J.F. Dodd, Use of ESR for the detection of irradiated dates (Phoenix dactylifera L.), Appl. Radiat. Isot. 47 (11–12) (1996) 1641–1645.
[26] H.M. Shahbaz, K. Akram, J.J. Ahn, J.H. Kwon, Investigation of radiation-induced free radicals and luminescence properties in fresh pomegranate fruits, J. Agr. Food Chem. 61 (17) (2013) 4019–4025.
[27] M.F. Desrosiers, W.L. McLaughlin, Examination of gamma-irradiated fruits and vegetables by electron spin resonance spectroscopy, Int. J. Radiat. Appl. Instrum. C Radiat. Phys. Chem. 34 (6) (1989) 895–898.
[28] E.F.O. de Jesus, A.M. Rossi, R.T. Lopes, Identification and dose determination using ESR measurements in the flesh of irradiated vegetable products, Radiat. Phys. Chem. 52 (5) (2000) 1375–1383.
[29] M. Cutrubinis, H. Delincée, M. Stahl, O. Röder, H.J. Schaller, Detection methods for cereal grains treated with low and high energy electrons, Radiat. Phys. Chem. 72 (5) (2005) 639–644.
[30] K. Malec-Czechowska, A.M. Strzelczak, W. Dancewicz, W. Stachowicz, H. Delincée, Detection of irradiation treatment in dried mushrooms by photostimulated luminescence, EPR spectroscopy and thermoluminescence measurements, Eur. Food. Res. Technol. 216 (2) (2003) 157–165.
[31] M. Ukai, Y. Shimoyama, Free radicals in irradiated pepper: an electron spin resonance study, Appl. Magn. Reson. 24 (1) (2003) 1–11.
[32] E. Bortolin, E. Bustos Griffin, E. Cruz-Zaragoza, V. De Coste, S. Onori, Electron paramagnetic resonance detection of Mexican irradiated spices, Int. J. Food. Sci. Tech 41 (4) (2006) 375–382.
[33] K. Akram, J.-J. Ahn, J.-H. Kwon, Characterization and identification of gamma-irradiated sauces by electron spin resonance spectroscopy using different sample pre-treatments, Food Chem. 138 (2–3) (2013) 1878–1883.
[34] S.K. Chauhan, R. Kumar, S. Nadanasabapathy, A.S. Bawa, Detection methods for irradiated foods, Compr. Rev. Food Sci. F. 8 (1) (2009) 4–16.
[35] N.D. Yordanov, V. Gancheva, M. Radicheva, B. Hristova, M. Guelev, O. Penchev, Comparative identification of irradiated herbs by the methods of electron paramagnetic resonance and thermoluminescence, Spectrochim. Acta A 54 (14) (1998) 2413–2419.
[36] H. Shields, P. Hamrick, X-irradiation damage of sucrose single crystal, J. Chem. Phys. 37 (1) (1962) 202–203.

[37] E. Sagstuen, E.O. Hole, Radiation produced radicals, in: M. Brustolon, E. Giamello, (Eds.), Electron Paramagnetic Resonance: A Practitioners Toolkit, John Wiley & Sons, New Jersey, 2009

[38] E. Georgieva, L. Pardi, G. Jeschke, D. Gatteschi, L. Sorace, N.D. Yordanov, High-field/ high-frequency EPR study on stable free radicals formed in sucrose by gamma-irradiation, Free Radic. Res. 40 (6) (2006) 553–563.

[39] G. Vanhaelewyn, E. Pauwels, F. Callens, M. Waroquier, E. Sagstuen, P. Matthys, Q-Band EPR and ENDOR of low temperature X-irradiated β-d-fructose single crystals, J. Phys. Chem. A 110 (6) (2006) 2147–2156.

[40] H. Vrielinck, J. Kusakovskij, G. Vanhaelewyn, P. Matthys, F. Callens, Understanding the dosimetric powder EPR spectrum of sucrose by identification of the stable radiation-induced radicals, Radiat. Prot. Dosim. 159 (1–4) (2014) 118–124.

[41] G.P. Guzik, W. Stachowicz, J. Michalik, Identification of irradiated dried fruits using EPR spectroscopy, Nukleonika 60 (3) (2015) 627–631.

[42] J.-H. Lee, J.-J. Ahn, K. Akram, G.-R. Kim, J.-H. Kwon, Comparison of electron spin resonance (ESR) spectra of irradiated standard materials using different ESR spectrometers, J. Korean Soc. Appl. Biol. Chem. 55 (3) (2012) 407–411.

[43] V. Bercu, C.D. Negut, O.G. Duliu, EPR investigation of some desiccated Ascomycota and Basidiomycota gamma-irradiated mushrooms, Radiat. Phys. Chem. 79 (12) (2010) 1275–1278.

[44] K. Akram, J.-J. Ahn, J.-H. Kwon, Identification and characterization of gamma-irradiated dried *Lentinus edodes* using ESR, SEM, and FTIR analyses, J. Food Sci. 77 (6) (2012) C690–C696.

[45] I.-D. Choi, B.-K. Kim, H.-P. Song, M.-W. Byun, M.-C. Kim, J.-O. Lee, et al., Irradiation detection in Korean traditional soybean-based fermented powdered sauces: data for establishing a database for regulation of irradiated foods, J. Korean Soc. Food Sci. Nutr. 10 (1) (2005) 29–33.

[46] N.D. Yordanov, O. Lagunov, K. Dimov, EPR spectra induced by gamma-irradiation of some dry medical herbs, Radiat. Phys. Chem. 78 (4) (2009) 277–280.

[47] R. Yamaoki, S. Kimura, K. Aoki, S. Nishimoto, Detection of electron beam irradiated crude drugs by electron spin resonance (ESR), Radioisotopes 56 (2007) 163–172.

[48] K.I. Aleksieva, K.G. Dimov, N.D. Yordanov, Identification of gamma-irradiated fruit juices by EPR spectroscopy, Radiat. Phys. Chem 103 (1) (2014) 27–30.

[49] B. Sanyal, S. Chawla, A. Sharma, An improved method to identify irradiated rice by EPR spectroscopy and thermoluminescence measurements, Food Chem 116 (2) (2009) 526–534.

[50] L. Calucci, C. Pinzino, M. Zandomeneghi, A. Capocchi, S. Ghiringhelli, F. Saviozzi, et al., Effects of γ-irradiation on the free radical and antioxidant contents in nine aromatic herbs and spices, J. Agric. Food Chem. 51 (4) (2003) 927–934.

[51] R.M. Slave, C.D. Negut, V.V. Grecu, ESR on some gamma-irradiated aromatic herbs, Rom. J. Phys. 59 (7–8) (2014) 826–833.

[52] V. Bercu, C.D. Negut, O.G. Duliu, EPR studies of the free radical kinetics in γ-rays irradiated *Pleurotus ostreatus* oyster mushrooms, Food Res. Int. 44 (4) (2011) 1008–1011.

[53] N.D. Yordanov, V. Gancheva, A new approach for extension of the identification period of irradiated cellulose-containing foodstuffs by EPR spectroscopy, Appl. Radiat. Isot. 52 (2) (2000) 195–198.

[54] EN 1788, Foodstuffs — Thermoluminescence Detection of Irradiated Food from which Silicate Minerals Can Be Isolated, 2001.

[55] N.D. Yordanov, K. Aleksieva, I. Mansour, Improvement of the EPR detection of irradiated dry plants using microwave saturation and thermal treatment, Radiat. Phys. Chem. 73 (1) (2005) 55–60.

[56] M. Polat, M. Korkmaz, Detection of irradiated black tea (*Camellia sinensis*) and rooibos tea (*Aspalathus linearis*) by ESR spectroscopy, Food Chem. 107 (2) (2008) 956–961.

[57] M.M. Sünnetçioğlu, D. Dadaylı, S. Çelik, H. Köksel, Detection of irradiated wheat using the Electron Paramagnetic Resonance Spin Probe Technique, Cereal Chem. 75 (6) (1998) 875–878.

[58] D. Dadaylı Paktaş, M.M. Sünnetçioğlu, EPR spin probe investigation of irradiated wheat, rice and sunflower seeds, Radiat. Phys. Chem. 76 (1) (2007) 46–54.

[59] E.F.O. de Jesus, A.M. Rossi, R.T. Lopes, An ESR study on identification of gamma-irradiated kiwi, papaya and tomato using fruit pulp, Int. J. Food Sci. Tech. 34 (2) (1999) 173–178.

[60] J.-Y. Kwak, K. Akram, J.-J. Ahn, J.-H. Kwon, ESR-based investigation of radiation-induced free radicals in fresh vegetables after different drying treatments, Int. J. Food Prop. 17 (6) (2014) 1185–1198.

[61] J.-J. Ahn, B. Sanyal, K. Akram, J.-H. Kwon, Alcoholic extraction enables EPR analysis to characterize radiation-induced cellulosic signals in spices, J. Agric. Food Chem. 62 (46) (2014) 11089–11098.

CHAPTER 2

ESR Investigation of the Free Radicals in Irradiated Foods

O.G. Duliu and V. Bercu

University of Bucharest, Magurele (Ilfov), Romania

Contents

2.1 INTRODUCTION

A short time after the discovery of X-rays in 1895 by Wilhelm Conrad Röntgen, and of natural radioactivity in 1886 by Henry Bequerel, Samuel Prescott [1] showed that radium radiation has a bactericide effect on some bacteria and yeasts, and David C. Gillett patented the first instrument for microbial decontamination of organic matter by using X-rays [2].

Beginning with these first applications, the technique of microbial decontamination by irradiation with high energy ionizing radiation steadily developed so, at present, it is currently used in various fields, such as the sterilization of single-use medical supplies [3] or food processing [4,5].

Numerous studies have showed that irradiation with high energy gamma rays or electrons appears to be the only method which inactivates pathogenic germs without affecting the raw food flavor or texture in a discernible manner [6–8]. Moreover, the macronutrients contained in food such as lipids, proteins, and carbohydrates, as well as the essential minerals, are not affected at all by this kind of treatment [9,10].

Electron Spin Resonance in Food Science.

17

Although the use of gamma rays, as well as high energy electrons, to decontaminate a large variety of foods is an accepted and well-known technique which now is utilized in more than 60 countries [11], consumers should be correctly informed about the fact that the food was decontaminated by irradiation. Therefore, in many countries, including EU members [12], there are certain regulations regarding the proper information which should be supplied to consumers about the use of ionizing radiation in the process of microbial food decontamination [13].

Currently there are several high sensitivity methods capable of detecting irradiation treatment applied to foods [14]. Among them, thermoluminescence [15,16] and especially electron spin resonance (ESR), due to its capacity to prove the presence of free radicals at contents of less than 10^{-6} ppm, has been shown to be the best suited in the case of irradiated hard tissue of all categories of food [17]. In recognizing its ability to prove the generation of paramagnetic free radicals upon gamma rays or electron beam irradiation of foods, the European Committee for Standardization (CEN) proposed ESR spectroscopy as a standard analytical technique for the detection of irradiated food containing cellulose [18], crystalline sugar [19], or bone [20].

As a rule, the ESR spectra of irradiated food consist of a reduced number of components, relatively easily identified by both g-factors and line shape. But, in some cases, such as the hard tissues of crustaceans, mollusk shell [21–23], cuttlefish phragmocone [24], or solid vegetables [25–27], the corresponding ESR spectra presented a relatively complex structure resulting from the superposition of the resonance lines of different free radicals (see Fig. 2.1).

In this case, the use of a higher ESR frequency, such as that corresponding to K (10.9–36.0 GHz) or Q (36.0–46.0 GHz) bands, could significantly increase the spectrometer resolution allowing separation of the resonance lines which overlap in the X-band (6.2–10.9 GHz) [28,29].

It is worth mentioning that ESR spectroscopy is the only analytical method which can directly detect the presence of free radicals, including those generated by ionizing radiation.

Important information regarding the free radicals induced by ionizing radiation in food stuffs is related to their shelf life, as well as their temperature stability under different kinds of thermal treatments. In the latter case, useful information could be obtained by means of two types of thermal treatments applied to irradiated hard tissues: (1) isothermal annealing, i.e., constant temperature and gradually increased annealing time [24,30–35],

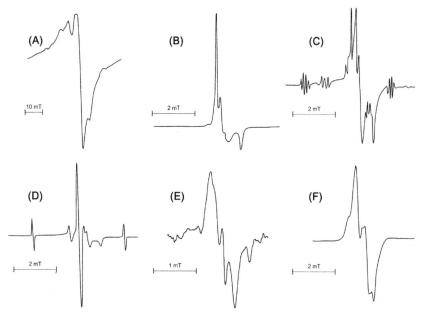

Figure 2.1 Some examples of intricate X-band ESR spectra reported for different irradiated hard tissue such as blue crab (*Callinectes sapidus*) (A)—after Maghrabi [23], oyster (*Crassostrea gigas*) (B)—after Della Monaca et al. [22], Venus shell (*Callista chione*) (C)— after Alberti et al. [21], cuttlefish (*Sepia officinalis*) phragmocone (D)—after Duliu [24], as well as dry ginger (*Zingiber officinale*) (E)—after Yamaoki et al. [27], and dry yarrow (*Achillea millefolium*) (F)—after Aleksieva et al. [25].

(2) isochronous processing, i.e., gradually increased temperature and constant annealing time [30,31,35,36].

Using these two different analyses, one can determine the half-life time ($T_{1/2}$) of free radicals from the decay of the ESR signal, as well as the activation energy obtained from the corresponding Arrhenius plots.

2.2 QUANTITATIVE ANALYSIS OF THE FREE RADICALS

The analysis of shelf life, as well as of the temperature stability of the free radicals, requires a quantitative analysis of the paramagnetic species. Because the ESR signal is proportional to the content of the paramagnetic centers, an absolute value of the free radical content can be obtained by analyzing the ESR line shape. However, this procedure is not straightforward (for a full analysis of quantitative ESR consult [37]). In this regard, there are two different approaches: (1) either using the free radical whose

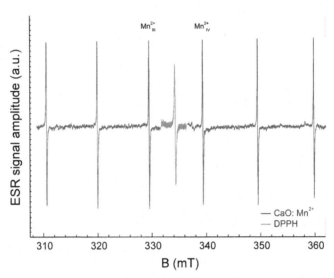

Figure 2.2 The experimental X-band ESR spectra of Mn^{2+} in CaO and the polycrystalline DPPH (original). The Mn^{2+} spectrum consists of six hyperfine structure lines with a distance between them of about 0.9 mT, corresponding to a g-factor of 2.0292 and 1.9760 for the third and fourth lines [38], i.e., those usually used for calibration.

g-factor is around two, or (2) using transition element ions whose ESR spectrum present a hyperfine structure.

From the first category, the solid DPPH (2,2-diphenyl-1-picrylhydrazyl) presents several advantages, such as a single narrow ESR line with a half-width of 0.5 mT and a g-factor of 2.0036.

In the second category, the Mn^{2+} spectrum is routinely used for quantitative measurements regarding the time, as well as temperature, evolution of the irradiation free radicals (see Fig. 2.2).

Indeed, a more appropriate category of markers for studies of the free radicals induced in food stuffs consists of the transition metal ions with a high value of hyperfine interaction, such as the Mn^{2+} ion with a well-known electron configuration, i.e., effective spin $S = 5/2$ and nuclear spin $I = 5/2$.

If present, the Mn^{2+} spectrum could be very well used as an internal marker for free radical ESR spectrum amplitude [39], especially because their lines are localized between the third and the fourth Mn^{2+} hyperfine structure lines (see Fig. 2.3).

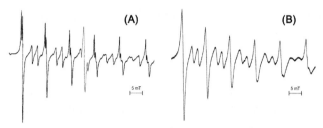

Figure 2.3 Some examples of *X*-band ESR Mn²⁺ ions in unirradiated mussel *Mytillus galoprovincialis* shell (A)—after Seletchi and Duliu [40]; used with permission. Copyright © 2007 Editura Academiei Romane, and freshwater crayfish *Astacus leptdactylus* cuticle (B)—after Bercu et al. [41]. Beside the Mn²⁺ sextet, the ESR spectrum of nonirradiated mussels presents a relatively complex free radical spectrum localized between the third and the fourth lines of the Mn²⁺ spectrum, represented in red (gray in print versions) (A).

The main advantage of the Mn²⁺ ion consists of its quasi-ubiquity, as it is present in almost all carbonates, regardless whether they are natural rock such as limestone, or constituents of an invertebrate exoskeleton [40].

Accordingly, Fig. 2.3 reproduces the experimental ESR spectra of Mn²⁺ in two unirradiated samples of the exoskeleton belonging to two completely different classes of animals well appreciated as food: *Mytillus galoprovincialis* mussels (Fig. 2.3A), and freshwater crayfish *Astacus leptodactylus* (Fig. 2.3B).

As one can see, the Mn²⁺ ESR spectra are spread over 70 mT with six lines due to the hyperfine structure, while the free radical spectra have one or more lines at a magnetic field close to the free electron *g*-factor value (Fig. 2.1 and Fig. 2.3A). In the case where the Mn²⁺ ions are not present in the sample, an external calibration sample could be used, preferably kept in the same position within the ESR spectrometer resonance cavity during measurements. As mentioned before, the Mn²⁺ ions in MgO or CaO are the most suitable calibration samples in the case of irradiation free radicals [42].

2.3 IRRADIATION FREE RADICALS TIME AND TEMPERATURE DEPENDENCE

2.3.1 Generation

As mentioned above, ^{60}Co gamma rays or accelerated electrons are currently used for microbial decontamination, the irradiation being performed at a constant dose rate until the programmed dose is reached. In this way, during irradiation, free radicals are generated at a constant rate and, at the same time, they recombine so the entire process could be described by the following differential equation:

$$\frac{dN}{dt} = N_0 p\dot{D} - \lambda N \tag{2.1}$$

where N represents the number of free radicals at time t; N_0 represents the number of bonds which could be broken; p represents the probability of a bond being broken by radiation on the unit of dose rate; $\dot{D}$ is the dose rate; and λ represents the probability of the recombination in the unit of time.

By solving the differential Eq. (2.1), considering that before irradiation the content of free radicals was zero, it follows that:

$$N = \frac{N_0 p\dot{D}}{\lambda}(1 - e^{-\lambda t}) \tag{2.2}$$

Further, by amplifying and dividing the exponent λt with the dose rate $\dot{D}$, Eq. (2.2) becomes:

$$N = N_{sat}\left(1 - e^{-\frac{D}{D_0}}\right) \tag{2.3}$$

where N_{sat} represents the number of free radicals in the sample when the absorbed dose tends to infinity; and D_0 represents the absorbed dose necessary for the number of free radicals to be e times lower than the saturation value N_{sat}.

As the amplitude of the ESR signal is proportional to the number of free radicals contained by the investigated sample, the ESR line amplitude represents a measure of the free radical content. This simplified model was experimentally confirmed by different authors [26,31,35,43–45], as the graphs reproduced in Figs. 2.4A–C. Moreover, Eq. (2.3) allowed us to calculate the numerical values of D_0, whose value depends on the nature of the irradiated media.

If the food was previously irradiated, as in the case of chicken bones [44], Eq. (2.3) was changed accordingly to include the original dose D_b:

$$N = N_{sat}\left(1 - e^{-\frac{D+D_b}{D_0}}\right) \tag{2.4}$$

In the case where D_0 is significantly higher than the dose usually used for microbial decontamination or sterilization, i.e., 10–15 kGy [48],

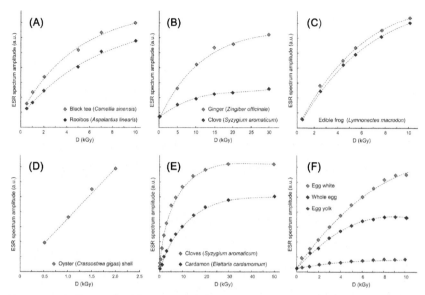

Figure 2.4 Different models proposed for dose dependency of the ESR spectrum amplitude. Exponential in the case of black tea (*Camellia sinensis*) and rooibos tea (*Aspalantus linearis*) (A)—after Polat and Korkmaz [31], clove (*Syzygium aromaticum*) and ginger (*Zingiber officinale*) (B)—after Polovka et al. [32]; used with permission. Copyright © 2007 VUP Food Research Institute, Bratislava, frog (*Limnonectes macrodon*) leg bone (C)—after Bercu et al. [35]; Linear or second order polynomial in the case of oyster (*Crassostrea gigas*) shell (D)—after Della Monaca et al. [22], cloves (*Syzygium aromaticum*) and cardamon (*Elettaria cardamomum*) (E)—after Beshir [46], and egg powder (F)—after Aydin [47].

Eq. (2.3) turns into a linear dependency of the absorbed dose, k, representing the number of free radicals generated by the unit of dose, i.e.:

$$N = N_0 + kD \tag{2.5}$$

as Fig. 2.5A illustrates.

Similar results were reported in the case of irradiated maize (*Zea mays*), for dose values between 0 and 11 kG [49], high water content papaya (*Carica papaya*) fruit for doses up to 3 kGy [50], irradiated oyster shell for doses up to 3 kGy [22], or different varieties of irradiated amylase and pectinase for doses up to 7 kGy [51].

For simple, unresolved ESR spectra, as in the case of egg powder, where the irradiated matter was characterized by a single resonance line, the dose dependency of the ESR line amplitude followed a more complex dependency. This would be better described by a second degree polynomial,

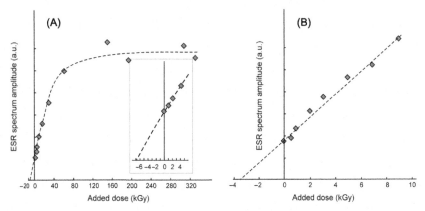

Figure 2.5 The additive dose techniques used to determine the original dose in the case of chicken bones (A)—after Desrosier [43] and (B)—after Parlato et al. [44]. In both cases, the original dose was less than 10 kGy. It should also be noted that at low doses (up to 6–10 kGy), the ESR signal amplitude follows the adsorbed dose by a linear dependency.

assuming the existence of different species of irradiation free radicals [47], or even by a third degree polynomial, as reported by Beshir [46].

Similar empirical dose dependencies were reported by Çam and Engin [30] for irradiated sage tea (*Salvia officinalis*) (second order polynomial), by Della Monaca et al. [22] in the case of irradiated oyster (*Crassostrea gigas*) shell, by Beshir [46] for cloves (*Syzygium aromaticum*) and cardamon (*Elettaria cardamomum*), or for irradiated egg powder by Aydin [47] (see Fig. 2.4D–F).

The existence of the above-mentioned phenomenological model, which describes with accuracy the accumulation of free radicals during irradiation, can be used to assess the initial dose received by decontaminated food. Indeed, by irradiating the same sample with well-determined doses, it is possible to determine the original dose by extrapolating to zero the amplitude or the intensity of the ESR resonance lines. This procedure, known as the additive dose method [43,44,52,53], allows determination of the initial dose with an accuracy up to 25% (the main source of uncertainties consisting of signal fading during storage; see Fig. 2.5).

2.3.2 Storage Fading

Once created, the content of irradiation free radicals either gradually increases by homolytic degradation of the compounds with weak covalent bounds [54], or disappears by reacting with another component of the surrounding media [54,55].

Regarding the hard tissues, postirradiation recombination was reported in the absolute majority of cases [9,17,32]. Moreover, the recombination processes were attributed to first order reactions, so that the content of free radicals decreases in time by following an exponential decay function [24,26,33,55], i.e.:

$$A(t) = A_0 + A e^{-\ln 2 \frac{t}{T_{1/2}}} \tag{2.6}$$

where $A(t)$ represents the amplitude of the ESR spectrum line immediately after irradiation stopped; A_0 represents the amplitude of the same line before irradiation; A represents the contribution of the irradiation free radicals to the ESR spectrum; and $T_{1/2}$ represents the interval of time necessary for the amplitude A to reduce to half (see Fig. 2.6A).

In some cases, as in those reported by Bercu et al. [29,35], the free radical ESR line represents a superposition of a few lines whose g-factors are very close, so that their ESR lines completely overlap.

In this case, the amplitude $A(t)$ is better described by a sum of exponential decay functions, i.e.:

$$A(t) = A_0 + A_1 e^{-\ln 2 \frac{t}{T_{11/2}}} + A_2 e^{-\ln 2 \frac{t}{T_{21/2}}} \tag{2.7}$$

This approach allows for the estimation of the number of different species of free radicals which contribute to the observed ESR line. For instance, as Aydin [47] showed, in the case of egg powder, i.e., egg yolk,

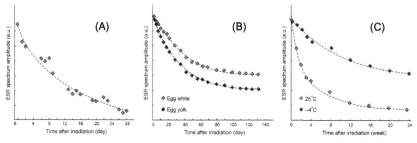

Figure 2.6 Time fading of the ESR signal of irradiation centers in *Pleurotus ostreatus* oyster mushroom following a simple exponential decay function (A)—after Bercu et al. [29], or a sum of the exponential decay function as in the case of egg (B)—after Aydin [47], as well as the temperature dependence of the ESR signals fading (C)—after Sudprasert et al. [56].

egg white, and whole egg, the time dependency of the irradiation free radicals could be described by a sum of two exponential decay functions, with different values of the A_0 and $T_{1/2}$ characteristic parameters (see Fig. 2.6B).

As expected, the storage temperature significantly influences ESR signal fading, so that the lower the temperature, the greater the presence of free radicals in irradiated foods (see Fig. 2.6C) [56].

The time stability of irradiation free radicals is strongly influenced by both the free radicals' nature, and the irradiated tissue types.

Accordingly, the most unstable radicals are those associated with irradiated cellulose, whose presence could be identified less than 70–90 days after gamma ray irradiation [57,58]. Conversely, free radicals induced in ginger powder [26], fresh water crayfish (*Astacul leptodactylus*) cuticle [41], or dehydrated oyster mushroom (*Pleurostus ostreatus*) [41] showed a remarkable stability, leading to their presence being demonstrated after 11, 14, and 16 months of ambient temperature storage, respectively (see Fig. 2.7). The same remarkable stability was also found in the case of CO^{2-} radicals in the hydroxylapatite of vertebrate bones [39].

Storage fading plays a significant role in the screening of any radiation microbial decontamination treatment. Accordingly, if the characteristic

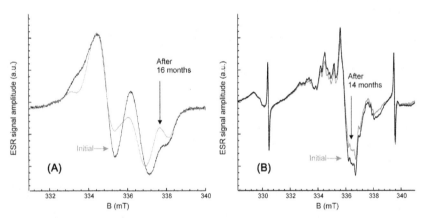

Figure 2.7 The ESR spectra of irradiated dehydrated oyster mushroom (*Pleurotus ostreatus*) (A)—Bercu et al. [41], and fresh water crayfish (*Astacul leptodactylus*) cuticle (B)—Bercu et al. [41], following ^{60}Co gamma ray irradiation and after 16 and 14 months of room temperature storage, respectively. In both cases, the storage fading appears to be less significant.

ESR signal is present, a previous ionizing radiation treatment could be presumed, but the absence of such a signal could be interpreted either as the absence of any radiation decontamination, or the total fading of the irradiation free radicals.

2.3.3 High Temperature Annealing

As previously mentioned, the ESR signal amplitude is proportional to the free radical content, so that it reflects the influence of the ambient temperature [59]. As a rule, during room temperature storage, recombination causes EPR signal amplitude to decrease, whereas at higher temperatures the generation of new species of free radicals is possible.

Therefore, an important amount of information about the free radicals' temperature behavior can be inferred from the temperature dependence of reaction rates [60].

In this regard, the reaction rate represents the parameter which better describes the evolution of the paramagnetic centers. According to [59], it can be defined as the change in the number of paramagnetic species in a unit of time, i.e., the change of the ESR signal amplitude.

Therefore, important information on the free radicals' temperature behavior can be inferred from the temperature dependence of reaction rates [60]. Moreover, as reaction rates could be regarded as authentic free radical fingerprints, the analysis of the temperature dependency of the corresponding ESR signals makes a better identification of the studied radicals possible. This approach is recommended when the ESR signals of different radicals overlap.

In this case, the temperature dependence of the reaction rate k is better described by the Arrhenius equation:

$$\ln k = \ln u - \frac{E_a}{RT} \tag{2.8}$$

or alternatively:

$$k = u e^{-\frac{E_a}{RT}} \tag{2.9}$$

where u is the pre-exponential factor; E_a is the activation energy; and R is the gas constant, equal to 8.314 J/K/mol.

Therefore, a plot of $\ln k$ versus $1/T$ will be linear, with a negative slope equal to E_a/R, thus permitting calculation of the E_a parameter.

If the reaction rate k is proportional to the relative amplitude of the ESR signal, it is defined as follows [35]:

$$I(T) = \frac{I_0 - I_T}{I_0} \qquad (2.10)$$

where I_0 is the reference amplitude of the ESR signal of the untreated sample; and I_T is the amplitude of an irradiated sample treated at different temperatures, all of them normalized to a reference sample.

Then the relative amplitude $I(T)$ of the ESR signal can be expressed in a similar logarithmic form:

$$\ln I(T) = \ln u - \frac{E_a}{RT} \qquad (2.11)$$

which makes the determination of the activation energy E_a possible (see Fig. 2.8).

Thus, in the case of irradiated *Limnonectes macrodon* frog bones (see Fig. 2.8A), the Arrhenius plot consists of two straight lines with different slopes, indicating the presence of two different irradiation free radicals with activation energies of $6.7 \pm 0.7\,\text{kJ/mol}$ (room temperature at $74°C$) and $7.3 \pm 0.6\,\text{kJ/mol}$ (above $74°C$), respectively [35].

A similar behavior is shown by the irradiation free radicals induced by gamma-rays in lyophilized *Pleurotus ostreatus* mushroom tissue (Fig. 2.8B). In this case, the corresponding Arrhenius plot displays three straight line segments, one of them with a negative slope corresponding to an activation energy of $42.64 \pm 2.85\,\text{kJ/mol}$, while the other two have positive

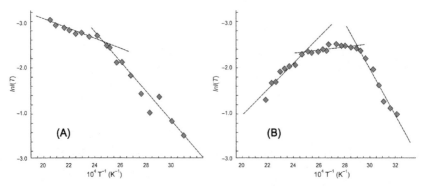

Figure 2.8 An Arrhenius plot of the relative amplitude of ESR signal of irradiated frog (*Lymnonectes macrodon*) bones—after Bercu et al. [35] (A); and oyster mushroom (*Pleurotus ostreatus*) lyophilized tissue—after Bercu et al. [29] (B).

slopes whose activation energies of $-2.74 \pm 1.41\,\text{kJ/mol}$ and $-23.94 \pm 2.57\,\text{kJ/mol}$ point toward a thermal generation of free radicals [29].

These observations emphasize the role of Arrhenius plots in identifying and describing the thermal behavior of irradiation free radicals, information which could not be obtained either by spectrum simulation, or by increasing the frequency of the ESR spectrometer.

2.4 CONCLUSIONS

ESR proved to be the preferred method to investigate irradiation free radicals in variety of foods. In view of this, a significant volume of literature data showed that ESR could be used not only to prove the presence of irradiation free radicals, but also to provide quantitative information regarding the kinetics of their generation or disappearance. As generation and fading could be described by first order reactions, the time evolution of irradiation radicals can be described by a single or linear combination of exponential decay, or exponential saturation functions, respectively.

This model is very helpful in proving the presence of different free radicals whose close values of g-factors determine overlapping of the characteristic ESR lines.

The simple saturation exponential model allows, by using consecutive irradiation with known doses of radiation with an uncertainty up to 25%, the determination of the initial dose received by the food during microbial decontamination by irradiation.

A more detailed analysis of the irradiation free radicals could be performed by controlled thermal annealing, either isothermal or isochronous. In the first case, the temperature is kept constant while the annealing time is gradually increased. In the second case, the temperature is gradually increased while the ESR spectrum is recorded at equal intervals.

Therefore, isothermal annealing could be assimilated with accelerated ageing, whereas isochronous annealing allows determination of the activation energy of each type of free radical present in the sample.

At the same time, the ESR should be considered an excellent screening method for identifying food decontaminated by irradiation, regardless of their nature or origin. In the case where a residual ESR signal is observed, one can assert a previous microbial decontamination by irradiation, but the absence of such a signal does not signify the absence of any irradiation treatment. In this case, complementary methods such as thermoluminescence could give a final answer.

REFERENCES

[1] S.C. Prescott, The effect of radium rays on the colon bacillus, the diphtheria bacillus and yeast, Science 503 (1904) 246–248.

[2] D.C. Gillett, Apparatus for preserving organic materials by the use of x-rays, US Patent No. 1,275,417, 1906.

[3] IAEA, Trends in Radiation Sterilization of Health Care Products, International Atomic Energy Agency, Vienna, 2008, ISBN:978-9201110077.

[4] G.J. Hallman, Phytosanitary applications of irradiation, Compr. Rev. Food Sci. Food Saf. 10 (2011) 143–151.

[5] R.A. Mollins (Ed.), Food Irradiation: Principles and Applications, Wiley-Interscience, New York, 2001. ISBN:978-0471356349.

[6] J. Farkas, Irradiation as a method for decontaminating food: a review, Int. J. Food. Microbiol. 44 (1998) 189–204.

[7] CGFI, Facts about food irradiation, International Consultative Group on Food Irradiation (ICGFI), Food and Environmental Protection Section Joint FAO/IAEA Division of Nuclear Techniques in Food and Agriculture International Atomic Energy Agency, Vienna, Austria. <http://www.piwet.pulawy.pl/irradiacja/factsaboutfoodirradiation.pdf>, 1999 (last accessed 15.12.15).

[8] WHO, High Dose Irradiation: Wholesomeness of Food Irradiated with Doses Above 10 kGy. WHO Technical Report Series 890, World Health Organization, Geneva, Switzerland, 1999, ISBN:924-1208902.

[9] Anonymous, Safety of irradiated food, risk assessment studies, report no. 37, Centre for Food Safety, Food and Environmental Hygiene Department, Hong Kong. <http://www.cfs.gov.hk/english/programme/programme_rafs/files/RA_37_Safety_of_Irradiated_Food_final.pdf>, 2009 (last accessed 15.12.15).

[10] WHO, Safety and Nutrition Adequacy of Irradiated Food, World Health Organization, Geneva, Switzerland, 1994.

[11] Eurofin, Irradiation testing for correct labeling you can trust. <http://www.eurofins.com/en/food-feed-testing/food-irradiation.aspx>, 2015 (last accessed 15.12. 15).

[12] EC, Report from the Commission to the European Parliament and the council on food ingredients treated with ionising radiation for the year 2012, Bruxelles <http://www.ipex.eu/IPEXL-WEB/dossier/document/COM20140052.do>, 2014 (last accessed 15.12.15).

[13] Anonymous, Codex Alimentarius Commission: general standard for the labelling of prepackaged foods: CODEX STAN 1–1985, 2010 upgrade <http://www.codexalimentarius.org/standards/list-of-standards/en/>, 2010 (last accessed 15.12.15).

[14] R. Stefanova, N.V. Vasilev, S.L. Spassov, Irradiation of food, current legislation framework, and detection of irradiated foods, Food Anal. Methods 3 (2010) 225–252.

[15] S. Kumari, S.K. Chauhan, R. Kumar, S. Nadanasabapathy, A.S. Bawa, Detection methods for irradiated foods, Compr. Rev. Food Sci. Food Saf. 8 (2009) 4–16.

[16] B.-K. Kim, C.-T. Kim, S.H. Park, J.-E. Lee, H.-S. Jeong, C.-Y. Kim, et al., Application of thermo-luminescence (TL) method for the identification of food mixtures containing irradiated ingredients, Food Anal. Methods 8 (2014) 718–727.

[17] B. Sanyal, A. Sharma, EPR and TL Techniques in Identification of Irradiated Food: Basics, Concepts, and Potential of the Techniques, LAP LAMBERT Academic Publishing (2012), ISBN:978-3659180712.

[18] EN 1787:2000, Detection of Irradiated Food Containing Cellulose. Method by ESR Spectroscopy, European Committee for Standardization, Brussels, 2000.

[19] EN 13708:2000, Foodstuffs – Detection of Irradiated Food Containing Crystalline Sugar. Method by ESR Spectroscopy, European Committee for Standardization, Brussels, 2001.

[20] EN 1787:2000, Detection of Irradiated Food Containing Bone. Method by ESR Spectroscopy, European Committee for Standardization, Brussels, 1996.

[21] A. Alberti, E. Chiaravalle, P. Fuochi, D. Macciantelli, M. Mangiacotti, G. Marchesani, et al., Irradiated bivalve mollusks: use of EPR spectroscopy for identification and dosimetry, Radiat. Phys. Chem. 80 (2011) 1363–1370.

[22] S. Della Monaca, P. Fattibene, C. Boniglia, R. Gargiulo, E. Bortolin, Identification of irradiated oysters by EPR measurements on shells, Radiat. Phys. Chem. 46 (2011) 816–821.

[23] A. Maghrabi, Identification of irradiated crab using EPR, Radiat. Meas. 42 (2007) 220–224.

[24] O.G. Duliu, Electron paramagnetic resonance identification of irradiated cuttlefish (*Sepia officinalis* L.), Appl. Radiat. Isotopes 52 (2000) 1385–1390.

[25] K. Aleksieva, O. Lagunov, K. Dimov, N.D. Yordanov, EPR study on non- and gamma-irradiated herbal pils, Radiat. Phys. Chem. 80 (2011) 767–770.

[26] O.G. Duliu, R. Georgescu, S.I. Ali, EPR investigation of some traditional oriental irradiated spices, Radiat. Phys. Chem. 76 (2007) 1031–1036.

[27] R. Yamaoki, S. Kimura, M. Ohta, Analysis of electron spin resonance spectra of irradiated gingers: organic radical components derived from carbohydrates, Radiat. Phys. Chem. 79 (2010) 41–423.

[28] A. Lund, M. Shiotani, S. Shimada, Applications of quantitative ESR, in: A. Lund, M. Shiotani, S. Shimada, (Eds.), Principles and Applications of ESR Spectroscopy, Springer, Netherlands, 2010, pp. 409–438. ISBN:978-1402053443.

[29] V. Bercu, C.D. Negut, O.G. Duliu, EPR studies of the free radical kinetics in γ-rays irradiated *Pleurotus ostreatus* oyster mushrooms, Food Res. Int. 44 (2011) 1008–1011.

[30] S.T. Çam, B. Engin, Identification of irradiated sage tea (*Salvia officinalis* L.) by ESR spectroscopy, Radiat. Phys. Chem. 79 (2010) 540–544.

[31] M. Polat, M. Korkmaz, Detection of irradiated black tea *Camellia sinensis* and rooibos tea *Aspalathus linearis* by ESR spectroscopy, Food. Chem. 107 (2008) 956–961.

[32] M. Polovka, B. Vlasta, P. Simko, EPR spectroscopy: a tool to characterize gamma-irradiated foods, J. Food Nutr. Res. 46 (2007) 75–83.

[33] D.E. Seletchi, O.G. Duliu, R. Georgescu, ESR studies of gamma-irradiated *Rapana venosa* (Gastropoda, Muricidae) shell, Radiat. Phys. Chem. 76 (2007) 1650–1652.

[34] N.D. Yordanov, K. Aleksieva, K. Mansour, Improvement of the EPR detection of irradiated dry plants using microwave saturation and thermal treatment, Radiat. Phys. Chem. 73 (2005) 55–60.

[35] V. Bercu, C.D. Negut, O.G. Duliu, Detection of irradiated frog *Limnonectes macrodon* leg bones by multifrequency EPR spectroscopy, Food. Chem. 135 (2012) 2313–2319.

[36] B. Engin, C. Aydas, M. Polat, Detection of gamma irradiated fig seeds by analysing electron spin resonance, Food. Chem. 126 (2011) 1877–1882.

[37] G.R. Eaton, S.S. Eaton, D.P. Barr, R.T. Weber, Quantitative EPR, Springer-Verlag, Wien, 2010, ISBN:978-3-211-92947-6.

[38] N.D. Yordanov, V. Gancheva, V.A. Pelova, Studies on some materials suitable for use as internal standards in high energy EPR dosimetry, J. Radioanal. Nucl. Chem. 240 (1999) 619–622.

[39] M. Ikeya, New Applications of Electron Spin Resonance: Dating, Dosimetry and Microscopy, World Scientific, Singapore, 1993, ISBN:981-0211996.

[40] D.E. Seletchi, O.G. Duliu, Comparative study of ESR spectra of carbonates, Rom. J. Phys. 72, 744–763 <http://www.nipne.ro/rjp/2007/_52/_5-6/0657/_0669.pdf>, 2007 (last accessed 23.12.15).

[41] V. Bercu, D.C. Negut, O.G. Duliu, Unpublished results, 2016.

[42] N.D. Yordanov, S. Lubenova, Effect of dielectric constants, sample container dimensions and frequency of magnetic field modulation on the quantitative EPR response, Anal. Chim. Acta 403 (2000) 305–313.

[43] M.F. Desrosiers, Estimation of the absorbed dose in radiation processed food-2. Test of the EPR response function by an exponential fitting analysis, Appl. Radiat. Isotopes 42 (1991) 617–619.

[44] A. Parlato, E. Calderaro, A. Bartolotta, M.C. D'Oca, M. Brai, M. Marrale, et al., Application of the ESR spectroscopy to estimate the original dose in irradiated chicken bone, Radiat. Phys. Chem. 76 (2007) 1466–1469.

[45] R.M. Slave, C.D. Negut, V.V. Grecu, ESR on some gamma-irradiated aromatic herbs, Rom. J. Phys. 59 (2014) 826–833.

[46] W.E. Beshir, Identification and dose assessment of irradiated cardamom and cloves by EPR spectrometry, Radiat. Phys. Chem. 96 (2014) 190–194.

[47] T. Aydin, Detection and original dose assessment of egg powders subjected to gamma irradiation by using ESR technique, Radiat. Phys. Chem. 114 (2015) 43–49.

[48] R. Yamaoki, S. Kimura, M. Ohta, Electron spin resonance spectral analysis of irradiated royal jelly, Food. Chem. 143 (2014) 479–483.

[49] D. Marcu, G. Damian, C. Cosma, V. Cristea, Gamma radiation effects on seed germination, growth and pigment content, and ESR study of induced free radicals in maize Zea mays, J. Biol. Phys. 39 (2013) 625–634.

[50] M. Kikuchi, Y. Shimoyama, M. Ukai, Y. Kobayash, ESR detection procedure of irradiated papaya containing high water content, Radiat. Phys. Chem. 80 (2011) 664–667.

[51] O.G. Duliu, M. Ferdes, O.S. Ferdes, EPR study of some irradiated food enzymes, J. Radioanal. Nucl. Chem. 260 (2004) 273–277.

[52] F. Bordi, P. Fattibene, S. Onori, M. Pantaloni, ESR dose assessment in irradiated chicken legs, Radiat. Phys. Chem. 43 (1994) 487–491.

[53] S.P. Chawla, P. Thomas, Identification of irradiated meat using electron spin resonance spectroscopy: results of blind trials, Int. J. Food Sci. Technol. 39 (2004) 653–660.

[54] G.C. Fritsch, T. Lopez, J.L. Rodriguez, Generation and recombination of free radicals in organic materials studied by electron spin resonance, J. Magn. Reson. 16 (1974) 48–55.

[55] M. Korkmaz, M. Polat, Radical kinetics and characterization of the free radicals in gamma irradiated red pepper, Radiat. Phys. Chem. 62 (2001) 411–421.

[56] W. Sudprasert, S. Monthonwattana, A. Vitittheeranon, Identification of irradiated rice noodles by electron spin resonance spectroscopy, Radiat. Meas. 47 (2012) 640–643.

[57] J. Raffi, N.D. Yordanov, S. Chabane, L. Donifi, V. Gancheva, S. Ivanova, Identification of irradiation treatment of aromatic herbs, spices and fruits by electron paramagnetic resonance and thermoluminescence, Spectrochim. Acta A 56 (2000) 409–416.

[58] J. Sadecka, Irradiation of spices – a review, Czechoslovak J. Food Sci. 25 (2007) 231–242.

[59] S.K. Upadhyay, Chemical Kinetics and Reaction Dynamics, Springer (2006), ISBN:1-40204546-8.

[60] P.W. Atkins, J. de Paula, Physical Chemistry for the Life Sciences, Oxford University Press (2006), ISBN:978-0-19-956428-6.

CHAPTER 3

Electron Spin Resonance Technique in the Quality Determination of Irradiated Foods

K.P. Mishra
Ex Bhabha Atomic Research Center, Mumbai, Maharashtra, India

Contents

3.1 INTRODUCTION

In the modern world, radiation technology is frequently employed to pre-serve food, reduce the risk of food-borne illnesses, prevent the spread of invasive pests, and delay or eliminate sprouting or ripening of fruit and crop products like onions, and potato. Irradiation of food can prevent the division of microorganisms by damaging their vital molecules such as bacteria and molds which cause food spoilage and slow down ripening or maturation of certain fruit and vegetables by affecting their physiological processes. The Food and Agriculture Organization (FAO) has estimated that worldwide about 25% of all food production is lost due to insects, bacteria, and rodents. Thus, radiation treatment of food offers an important method for reducing losses and lowering dependence on chemical pesticides. Evidently, reducing the post-harvest spoilage of foods or increasing the shelf life of crop products can significantly help achieve food security and food safety.

Processing of food by atomic radiation consists of the controlled application of energy by exposure to gamma rays, electrons, or X-rays. The process involves exposing the food for a desired period, either packaged or in bulk, to highly penetrating ionizing radiation such as gamma rays and X-rays. Research on food irradiation dates back to the beginning of the last century, when the United States and Britain obtained patents for the use of ionizing radiation to kill bacteria in food in 1905. Today, authorities concerned with health and safety in over 60 countries have approved the irradiation of foods ranging from spices to grains, to deboned chicken meat, beef, fruit, and vegetables. But the major push to irradiate food came in 1983 after the standard was adopted by the Codex Alimentarius Commission, a joint body of the FAO and the World Health Organization (WHO), which determines food standards for protecting consumer health and facilitates proper food trade. The Codex General Standard for food irradiation was based on the findings of a Joint Expert Committee on Food Irradiation (JECFI) convened by the FAO, WHO, and the International Atomic Energy Agency (IAEA), who recommended that irradiation of foods up to 1 kGy did not require any toxicological test [1–3]. In practice, radioisotopes commonly used for irradiation are ^{60}cobalt and ^{137}cesium for gamma rays, whereas electrical machines are employed to generate X-rays and accelerate electrons. The food irradiation process involves passing the material through a radiation field, thereby allowing the food to absorb external radiation energy. The food itself never comes in contact with the radioactive source. The advantage of food sterilization by ionizing radiation consists in insignificant changes in the freshness and texture of food, unlike other methods of food processing. Treatment of foods by radiation delays the ripening of fruits and the sprouting of vegetables by slowing down the rate of enzymes produced in them. Research has shown that irradiation of food produces few or insignificant chemical changes, which are in any case not known to be harmful.

Radiation treatment is known to produce radiolytic species and free radicals which are suspected of causing health effects. A number of specific tests have been developed for testing and examination of irradiated foods for public consumption. However, no single method has yet been

Table 3.1 Standard methods for the detection of irradiated foods adopted by the European Committee for Standardization (CEN), 1999

Foodstuffs—Detection of irradiated food containing fat—Gas chromatographic analysis of hydrocarbons (EN 1784)

Foodstuffs—Detection of irradiated food containing fat—Gas chromatographic/ Mass spectrometric analysis of 2-alkylcyclobutanones (EN 1785)

Foodstuffs—Detection of irradiated food containing bone—Method by ESR spectroscopy (EN 1786)

Foodstuffs—Detection of irradiated food containing cellulose—Method by ESR spectroscopy (EN 1787)

Foodstuffs—Detection of irradiated food from which silicate minerals can be isolated—Thermoluminescence (EN 1788)

Source: Reproduced from "Facts about Food Irradiation," Report of International Consultative Group on Food Irradiation, FOA/IAEA Division of Nuclear Techniques in Food and Agriculture, 1999.

developed that reliably detects irradiated foods for the radiation doses commonly employed [4]. Electron spin resonance (ESR) spectroscopy and thermoluminescence methods are generally used for detecting radiation-induced signals in processed spices and meat containing bone tissue [1]. Studies have shown that untreated foods yield ESR signals which are attributable to paramagnetic ions ($g \sim 4.00$) and organic free radicals ($g \sim 2.0$). The ESR method also distinguishes between unirradiated and irradiated foods to a limited extent [5]. In view of technological progress, the volume and variety of radiation processed foods have grown steadily in many countries over the years. Based on research results and for the purpose of regulations (Table 3.1), generally the radiation doses employed are categorized into low (up to 1 kGy), medium (1–10 kGy), and high dose ranges (>10 kGy) for practical purposes [3,6,7].

3.2 RADIATION-INDUCED FREE RADICALS

Exposure of solid foods to radiation produces free radicals which are trapped in the matrix. It is necessary to detect and estimate these radicals for scientific purposes, and for quality evaluation. The ESR technique offers a unique and reliable method to detect free radicals and paramagnetic ions in untreated, as well as in radiation treated, solid foods [1]. The presence of free radicals in irradiated foods is a common scientific and public concern which needs to be adequately addressed. The ESR method seems to be the method of choice in quality evaluation of irradiated foods. However, it needs to be noted that small quantities of chemical species found in irradiated food are also found in foods processed by other methods, such as heat treatment. The major challenge is to detect and characterize free radicals and measure radical yields as a function of radiation doses. It is thus possible from the extrapolation method to determine the radiation dose that produces minimum amounts of free radical species in irradiated foods. It is important to note that the quantities of species formed in irradiated foods are mostly lower or similar to that formed by other food treatment processes. The radiation doses which cause toxic changes are much higher than the doses usually needed to accomplish the benefits of irradiation. Studies carried out on a variety of foods have shown that there is no significant risk from radiation-generated free radical species.

3.3 BASIC RADIATION CHEMICAL MECHANISMS

From the theory of radiation effects on solid materials, it is known that irradiation causes the generation of free radicals in food materials by direct as well as by indirect action on the molecules of the food items. Thus, radiation processing is expected to produce some changes in food chemistry. Radiation sources like gamma rays, and charged particles like high energy electrons, penetrate deep into the target material and, in the process, collide with the atoms and molecules of the target material and transfer their energy. As a consequence of the absorption of passing radiation energy, the electrons of the target atoms are stripped off (ionization process), or chemical bonds are broken producing short-lived radicals such as the hydroxyl radical (OH), hydrogen atoms (H), and electrons (e^-). These radicals cause further chemical changes by reacting with neighboring vital molecules. Radiation can cause damage either directly, or through the reactions of radiolytic radicals with the DNA, RNA, proteins, and lipids in food material. This mechanism of radiation action on pathogens inhibits their division. Irradiation thus causes a multitude of chemical changes by the production of radiation related products which are unique, but are not considered dangerous. However, the scale of these chemical changes is not unique. Cooking, smoking, salting, and other traditional techniques are known to cause alteration in foods so that sometimes its original nature is almost unrecognizable. Moreover, it is common knowledge in the food industry that storage of food material causes dramatic chemical changes that eventually lead to unacceptable quality deterioration and spoilage.

3.4 EVALUATION OF IRRADIATED FOODS

Policy makers, technologists, and researchers had recognized the importance of methods of detection of irradiated foods to achieve wider public acceptance and increased export trade in radiation processed foods early in 1988. The greater need for reliable and routine tests to determine whether or not food had been irradiated arose as a result of rapid progress in the commercialization of the food irradiation process. It has become a pressing issue to develop easy and simple methods of testing in view of the growing international trade in irradiated foods, the adoption of regulations in different countries for the use of the technology and, above all, increasing consumer demand for identifiable categorization of radiation treated foods. For greater acceptance of irradiated foods by consumers, developing reliable identification methods is necessary for regulation control. In fact, the methods can be supplemented by additional means of enforcement, thus facilitating international trade and reinforcing consumer confidence in the overall process.

3.4.1 Food Quality Issues

Because of certain radiation chemical reactions, some changes in food quality after irradiation are inevitable. The nutritional content of food, as well as the sensory qualities (taste, appearance, and texture) is affected by radiation treatment. The changes in quality and nutrition are found to vary greatly from food to food. The changes in the flavor of fatty foods like meats, nuts, and oils are sometimes easily noticeable, but the changes in lean products like fruit and vegetables are difficult to recognize. Some studies by the food processing industry show that consumers find improved sensory qualities of some properly treated fruit and vegetables following irradiation of the product compared to untreated fruit and vegetables. However, it remains to be confirmed on the basis of results using physical techniques and biochemical studies.

3.4.2 Methods for Identification of Irradiated Food

It has been seen that while irradiation kills pathogens, it can induce free radicals that may threaten human health. The ESR spectroscopy technique has been applied for the evaluation of irradiation treatment to a wide variety of foods [3,4,8–11]. The ESR technique is specific, rapid, and simple to use to detect radicals contained in foods. The greatest advantage of ESR is that it is a nondestructive testing method. Thus, it allows the food samples to be re-analyzed and results to be reconfirmed. In fact, ESR has been accepted as a standard method (by the European Committee for Standardization, CEN) in the EU community. In the past years, ESR has become increasingly popular in food control laboratories all over the world. Extensive research has been carried out since the 1980s, resulting in the development of a range of tests which can be used to reliably determine the irradiation status of a wide variety of foods. Among them, the methods that have been studied most extensively and that offer the greatest scope for applications are ESR spectroscopy and thermoluminescence (TL). These methods have been successfully evaluated in a number of inter-laboratory blind trials with the result that five tests have been adopted as standard reference methods for the detection of irradiated food (see Table 3.1) by the European Committee for Standardization (CEN) in 1996 [3,4,6]. These, in turn, have been adopted by some national regulatory authorities. Further research continues to develop improved tests for implementation as reference methods.

3.5 ELECTRON SPIN RESONANCE SPECTROSCOPY AND FREE RADICAL RESEARCH

3.5.1 Basic Principles

In principle, ESR detects paramagnetic centers with unpaired electrons called free radicals that may be either intrinsic to the sample, or radiation-induced. A strong external magnetic field generates a difference between the degenerate energy levels of the electron spins in terms of Zeeman splitting, $m_s = +\frac{1}{2}$ and $m_s = -\frac{1}{2}$, which results in resonance absorption of an applied microwave energy (~GHz), resulting in spin flipping. The major components of an ESR spectrometer are: an electromagnet (variable magnetic field); a sample cavity connected to a modulator of field; a Klystron (frequency generator); a crystal detector; a phase sensitive detector; and a recorder for recording the signal.

The absorption spectrum is obtained by measuring the intensity of absorption by the sample placed in an external magnetic field which is swept within a narrow range on application of a fixed incident microwave frequency (mostly in the X-band region). The peak of the absorption is obtained when the resonance condition is satisfied by matching the frequency of the incident microwave with the energy separation of spin states ($w = h\nu/gBH$). For the sake of precise measurements, the ESR spectrum is recorded as the first derivative of the absorption peak. A typical ESR spectrum is illustrated in Fig. 3.1.

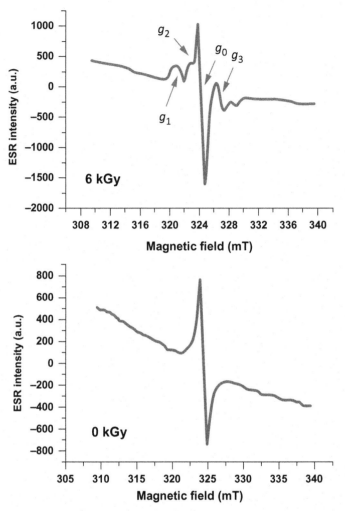

Figure 3.1 Illustration of a typical ESR spectra of irradiated broccoli sprout seeds after nitric acid (5%) extraction. *Adapted from J.-H. Kwon, H.M. Shahbaz, J.-J. Ahn, Advanced Electron Paramagnetic Resonance Spectroscopy for the Identification of Irradiated Food, Internet, January, 2014; J.H. Lee, J.J. Ahn, et al., Comparison of ESR spectra of irradiated standard materials using different ESR spectrometers. J. Korean Soc. Appl. Biol. Chem. 55 (2012) 407–411, Fig. 5, Internet Post, 2014.*

The absorption peak is measured at the cross-section of the derivative spectrum as a function of the applied magnetic field (EN 1787, 2000). The detected radical is identified by the g value (gyromagnetic ratio) of the ESR signal, which is the ratio of frequency to magnetic field strength ($h\nu/\beta H_0$, where h is Planck's constant, ν is the microwave frequency, β is the Bohr Magneton, and H_0 is the magnetic field). The g value and signal of the organic free radical of unirradiated and irradiated signals are the same, but their intensities differ and, notably, the complexity of spectra from irradiated samples varies. Radiation-induced paramagnetic species can remain stable in the rigid and dehydrated parts of a food sample for a longer time in comparison to the shelf life of the processed food.

3.5.2 Features of Electron Spin Resonance Spectra

ESR experiments are carried out by placing the sample in a quartz ESR tube (a few mm diameter, supplied by the manufacturer); the tube is sealed with paraffin film and stored in the dark in a desiccator. The sample is exposed to the simultaneous action of a magnetic field ($\sim$3.3 kG) and an electromagnetic microwave of very high frequency ($\sim$9.0 GHz). The ESR signals are measured as described in the European Standard Protocols. A number of research studies have reported radiation-induced signals in plant materials belonging to cellulose, crystalline sugar, and hypoapatite. Irradiated foods give characteristic ESR signals of cellulose with two side peaks having g values of $g = 2.0201$ and $g = 1.9851$, linked with a central signal (Fig. 3.1).

In contrast, irradiated sugar standard markers show the multi-component unparalleled radiation-induced signals that are characteristic of the hydroxy-alkyl radical in crystalline carbohydrates, as reported in irradiated food samples (rice noodles, irradiated wheat, dried mushrooms, and dried fruit [5]). It is further reported that the nature of paramagnetic centers in marine carbonates CO_2^{-1}, CO_3^{-3}, and CO_3^{-1}, are key carbonate-derived radicals used to define typical radiation-induced spectra. In particular, generation of CO_2^{-1} from carbonated materials upon irradiation is important, with a value of $g = 1.9996$, which is more specific than other radicals [5,8–10].

3.5.3 Electron Spin Resonance Signals in Irradiated Foods

In food materials of plant origin, radiation exposure can generate free radicals in cellulose and crystalline sugars, which may serve as irradiation detection markers in ESR analysis. Radiation-induced radicals generally occur in the solid and dry fractions of food. The rigid structure of the matrix is able to trap free radicals or excited states of the electrons and inhibits them to react with each other or with the food components present in the softer portions. Foods containing bone, seeds, shells, etc., have a low moisture content, and radicals remain sufficiently stable for ESR measurement and analysis. In general, it is difficult to apply the ESR technique to foods with a high moisture content, since free radicals produced during the irradiation process disappear rather rapidly. ESR spectroscopy is also utilized for the identification of irradiated fruit and vegetables. In high moisture products like fruit and vegetables, the irradiation-induced radicals are not stable. However, hard parts with low moisture where the free radicals are relatively stable can be utilized to detect irradiation treatment.

3.6 ELECTRON SPIN RESONANCE METHOD IN QUALITY ASSESSMENT

ESR is a specific, quick, and user-friendly research and application tool which can be employed for quantitative free radical estimation. It is a nondestructive technique that can detect paramagnetic ions and free radicals in a variety of normal and irradiated solid materials. The need for simple, reliable, and routine tests to determine whether or not food has been irradiated arose as a result of the progress made in the commercialization of the food irradiation process, greater international trade in irradiated foods, differing regulations relating to the use of the technology in many countries, and consumer demand for clear labeling of treated food in the market set up. Although not essential for management of the process, it was envisaged that the availability of such tests would help strengthen national regulations on the irradiation of specific foods, and enhance consumer confidence in such regulations. The availability of reliable identification methods would be of assistance in establishing a system of legislative control, and would help to achieve wider acceptance of irradiated foods by the public. It is recommended by experts that governments should undertake research into methods of detection of irradiated foods to ensure high quality and market confidence. Good Manufacturing Practice (GMP) should be followed in the preparation of food, including processing by irradiation or any other means [3,4,6].

3.7 INDIAN FOOD IRRADIATION PROGRAM AND ELECTRON SPIN RESONANCE RESEARCH

The Department of Atomic Energy, of the government of India has recognized the importance of food irradiation program for preserving and prolonging the shelf life of perishable foods by starting an extensive research and development program in the early years of the nuclear research program. Strong research backing exists at the Bhabha Atomic Research Center of the Department of Atomic Energy at Trombay, Mumbai, for radiobiological and toxicological studies of irradiated foods. Facilities for radiation treatment at a pilot scale in the laboratory at BARC, together with a research program on toxicological and biochemical characterization, are in place. Intensive and constant research backing for ensuring the safety of irradiated foods are the primary focus of the food processing technology program. Apart from many advanced research facilities and equipment for the characterization of irradiated foods, an X-band ESR spectrometer (Bruker Biospin AG, Switzerland) is employed for basic research and quality assessment of irradiated foods. The research group of the present author has investigated in considerable detail the mechanisms of radiobiological damage by ESR techniques in several biological systems [11–14]. Further research continues for specific applications and newer items are being evaluated for commercial and export purposes.

In 1994, the government of India approved the irradiation of onions, potatoes, and spices for international marketing and domestic public consumption. Thereafter, the department constructed two demonstration facilities for food irradiation. A facility for technology demonstration purposes is in place for the irradiation of spices at Vashi, Navi Mumbai. Another demonstration facility for irradiation of potatoes and onions is operational at Lasalgaon in the Nashik District of Maharashtra State. Based on the increasing confidence of consumers in technology backed demonstrations of success in the preservation of onions and potatoes, a few commercial enterprises have been licensed for wider utilization of radiation food processing technology. The program is strictly regulated by the Atomic Energy Regulatory Board (AERB) following the IAEA/WHO guidelines for the safety of irradiated foods. In order to provide consumers with a year-round supply of potato tubers, onion bulbs, yams, and other sprouting plant foods, storage over many months is necessary. Such long-term storage is possible with the aid of refrigeration, but it is costly, particularly in subtropical and tropical regions. A very low radiation dose of ~1 kGy or less inhibits sprouting of products such as potatoes, yams, onions, garlic, ginger, and chestnuts. It leaves no residues and allows storage at higher temperatures. Irradiation of potatoes stored at regulated temperatures (10–15°C) provides better processing quality. Irradiation technology is also available for sterilization of spices meant for local consumption, as well as for export. It is instructive to recall that most spices become heavily contaminated with microbes, including pathogenic bacteria, during sun drying. Contamination arises from the deposition of excreta of insects, birds, rodents, and other animals, and from wind-blown dust containing many microbes. Spices can be saved from caking and spoilage after radiation treatment. Radiation processed food cannot be identified by sight, smell, taste, or touch. Regulations require packages of irradiated foods to be marketed in India labeled with a specified logo, along with the written words indicating that they are processed by an irradiation method, with the date of irradiation, license number of the facility, and the purpose of irradiation. Consumers are well informed, and have a free choice whether to buy irradiated commodities or not [6].

3.8 FUTURE SCOPE

ESR has become an increasingly popular method for the identification of irradiated foods all over the world, and efforts are being made by researchers to extend the application of ESR methodology to detect and identify many types of irradiated foods. However, ESR results are affected by the nature of food and its water content. The power saturation behavior of intrinsic organic free radicals and that of radiation-generated radicals are found to be different; this needs to be utilized for quantitative quality assessment of radiation processed foods [10]. Recent reported results by Lee et al. suggest that the ESR spectrometer make dependent spectra of radicals in irradiated foods, which needs further investigation and confirmation [15]. The ESR spin labeling technique can be employed to detect and quantify free radical species which can sense its environment from normal to irradiated samples [16]. The technique has already been used successfully for herbs, nuts, spices, and meat. The technical need is to develop a table top X-band instrument with compensating circuits for loss of power in water containing samples. The development of a multi-band ESR and its utilization for detection and quantification of radicals may be desirable for newer applications in the future.

ACKNOWLEDGMENT

I would like to thankfully acknowledge the editorial and technical assistance of my son, Anil Mishra, from time to time. The silent support and enormous patience of my wife, Usha, and daughter-in-law, Mamta, during preparation of this chapter are duly acknowledged.

The research work of my co-authors as well as some illustrations taken from other researchers are duly acknowledged.

REFERENCES

[1] C.P. Poole, Jr., Electron Spin Resonance, Wiley-Interscience, Canada, 1983.

[2] Joint FAO/IAEA/WHO Expert Committee, Wholesomeness of Irradiated Food, WHO Geneva No. 659, 1981.

[3] J. Silverman, Current status of radiation processing, Radiat. Phys. Chem. 14 (1979) 17–21.

[4] E.M. Stewart, Detection methods for irradiated foods (Chapter 14), in: R.A. Molins (Ed.), Food Irradiation: Principles and Applications, John Wiley & Sons, New York, NY, 2001, pp. 347–386.

[5] J.-H. Kwon, H.M. Shahbaz, J.-J. Ahn, Advanced Electron Paramagnetic Resonance Spectroscopy for the Identification of Irradiated Food, Internet, January, 2014.

[6] Report of International Consultative Group on Food Irradiation, FOA/IAEA Division of Nuclear Techniques in Food and Agriculture, 1999.

[7] K.P. Mishra (Ed.), Biological Response, Monitoring and Protection from Radiation Exposure, Nova Science Publishers, Inc, New York, NY, 2015.

[8] M. Ukai, et al., J. Food Sci. 68 (2003) 2225–2229.

[9] M. Ukai, et al., Appl. Magn. Reson. 25 (2003) 95–103.

[10] M. Ukai, ESR studies on food irradiation research, Zeol News 39 (1) (2004) 24–27.

[11] K.P. Mishra, Radiation oxidative damage and program cell death in mammalian cells in vitro: fluorescence and EPR studies, Recent Tren. Biophys. Res. (2003) 87–89.

[12] K.P. Mishra, Radiation induced permeability changes in liposome and thymocytes: potential and relevance to biological dosimetry, Adv. ESR Appl. 18 (2002) 243–246.

[13] B.N. Pandey, K.P. Mishra, Fluorescence and ESR studies on membrane oxidative damage by gamma radiation, Appl. Magn. Reson. 18 (2000) 483–492.

[14] K.P. Mishra, B.B. Singh, A.R. Gopal-Ayengar, ESR studies on gamma irradiated TAN (triacetoneamine N-oxyl) radicals, Int. J. Radiat. Biol. 24 (4) (1973) 417–420.

[15] J.H. Lee, J.J. Ahn, et al., Comparison of ESR spectra of irradiated standard materials using different ESR spectrometers, J. Korean Soc. Appl. Biol. Chem. 55 (2012) 407–411.

[16] Y.T. Zhou, J.J. Yin, Y.M. Lo, Application of ESR spin label oximetry in food science, Magn. Reson. Chem. 49 (2011) 105–112.

CHAPTER 4

ESR Detection of Irradiated Food Materials

G.P. Guzik[1] and A.K. Shukla[2]
[1]Institute of Nuclear Chemistry and Technology, Warsaw, Poland
[2]Ewing Christian College, Allahabad, Uttar Pradesh, India

Contents

4.1 INTRODUCTION

Fresh food items normally remain fresh for a short time. Food irradiation is a popular way to achieve a longer storage time while maintaining the nutritional profile. Free radicals may be generated as a result of irradiation due to the breaking of bonds between molecules. Detection of irradiated food using the electron spin resonance (ESR) spectroscopy technique is therefore suitable, as it is applicable to paramagnetic samples (with unpaired electron spins). The non-destructive nature of ESR adds to its suitability for this purpose. In the quest for a reliable identification method for detection of irradiated foodstuffs worldwide, the European Union standardized three detection methods based on ESR for irradiated food containing bone (EN 1786: 1996), cellulose (EN 1787: 2000), and crystalline sugar (EN 13708: 2001) [1–3]. The theory of ESR and its applications have been described in many books and review articles. This chapter intends to introduce the applications of ESR spectroscopy in the study of irradiated/processed fruit, spices, herbs, and beverages.

Electron Spin Resonance in Food Science.

4.2 FRUIT

Initial ESR studies on irradiated fruits have been reported decades back in the literature. In 1971, Boshard et al. reported the ESR of irradiated papaya, and indicated the need to correlate the free radical content of fruit, irradiated and otherwise [4]. Some kinds of fruit, such as strawberries, contain manganese. Dodd et al. reported (Fig. 4.1) ESR spectra of excised strawberry achenes before and after electron irradiation (10 kGy dose) [5]. It shows a free radical signal and a Mn^{2+} signal. The free radical signal increases on irradiation, while the Mn^{2+} signal remains unchanged. Radicals induced by radiation were further found to be sufficiently stable to detect irradiation. They could suggest that the ratio of free radical and Mn^{2+} signals are related to the ripeness of fruit and the storage conditions.

There are many reports in literature in the following years on radiation–induced radicals that are trapped in dried fruit and survive in there for a relatively long time. The ESR technique has been suitably applied for the detection of radiation treatment of such dried fruit [6–10]. Raffi et al. [7] have reported their ESR investigations of strawberry, raspberry, red currant, bilberry, apple, pear, fig, french prune, kiwifruit, water-melon, and cherry. Fig. 4.2 shows the ESR spectrum of irradiated raspberry achenes. The authors could infer a weak triplet due to a cellulose radical just after γ irradiation.

A schematic ESR spectrum of an irradiated fruit (where A and B are absent) proposed by Raffi et al. [7] is shown in Fig. 4.3. This figure serves

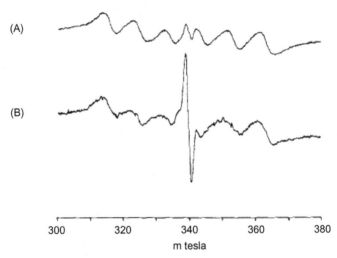

Figure 4.1 Electron Spin Resonance spectra of excised strawberry achenes: (A) before, and (B) after, electron irradiation (10 kGy dose). *Reproduced with permission from N.J.F. Dodd, A.J. Swallow, F.J. Ley, Use of ESR to identify irradiated food, Radiat. Phys. Chem. 26 (1985) 451–453.*

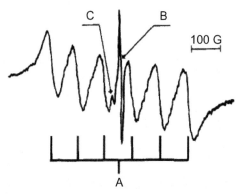

Figure 4.2 The ESR spectrum of irradiated raspberry achenes. Here A, present in both irradiated and nonirradiated samples, is assigned a six line signal due to Mn^{2+} [5]; B, also present in irradiated and nonirradiated samples, is a single line, the origin of which is unknown. Its height increases with the irradiation dose [5], but varies widely with the water content and consequently, cannot be used as proof of irradiation [6]. C only present in irradiated samples, is not a single line: it is linked to another one C′, which is not always visible at the right of the B signal because it is obscured by one of the Mn lines when A lines are present. B is almost at the center of the C–C′ doublet, therefore C and C′ may be part of a triplet. *Reproduced with permission from J. Raffi, J.-P. Agnel, Electron spin resonance identification of irradiated fruits, Radiat. Phys. Chem. 34 (1989) 891–894.*

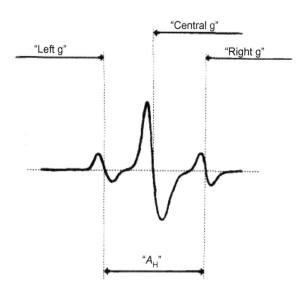

Figure 4.3 A schematic ESR spectrum of an irradiated fruit where A and B are absent, in order to define the ESR parameters: the central *g*-factor, which is in fact the *g*-factor of the B signal, and the "left" and "right" *g*-factors, which are the *g*-factors relative to the centers of the two lines of the C signal (considering them as independent). The hyperfine coupling constant A, i.e., the difference (in gauss) between the two lines of C signal, C being considered here as a doublet; if C is a triplet, A, will measure twice the value of the real hyperfine constant. *Reproduced with permission from J. Raffi, J.-P. Agnel, Electron spin resonance identification of irradiated fruits, Radiat. Phys. Chem. 34 (1989) 891–894.*

Table 4.1 ESR constants of irradiated fruits

Fruit	Central g	Left g	Right g	A_H	T_{max} (days)
Strawberries[a]	2.0043	2.0201	1.9852	60.4	23
Raspberries[a]	2.0038	2.0200	1.9852	61.0	20
Red currants[a]	2.0032	2.0199	1.9862	59.4	>18
Bilberries[a]	2.0032	2.0196	1.9855	60.6	>19
Mulberries[a]	2.0044	2.0201	1.9851	61.1	0
Apples[b,c]	2.0037	2.0197	1.9853	59.9	0
Pears[b,c]	2.0036	2.0198	1.9848	61.1	0
Figs[a]	2.0042	2.0204	1.9851	61.8	<19
Kiwifruit[b]	2.0044	2.0198	1.9858	59.4	0
Melon[b]	2.0044	2.0208	1.9856	61.4	0
Water-melon[b]	2.0039	2.0204	1.9845	62.6	0
French prunes[d]	2.0044	2.0204	1.9856	60.6	>15
Cherries[d]	2.0040	2.0201	1.9852	60.9	0
Coconuts[e]	2.0044	2.0211	1.9856	61.9	0
Average value	2.0040 ± 5	2.0202 ± 4	1.9853 ± 4	60.8 ± 0.9	

Source: Reproduced with permission from J. Raffi, J.-P. Agnel, Electron spin resonance identification of irradiated fruits, Radiat. Phys. Chem. 34 (1989) 891–894.
g-Factors determined on:
[a]Achenes.
[b]Pips.
[c]Stalks.
[d]Stones.
[e]Fibers.
T_{max}, maximum time of observation of the "doublet," in days.

to define the ESR parameters. B and C signals exist in all fruit studied by Raffi et al. [7]. A summary of ESR parameters is mentioned in Table 4.1. The C signal, due to its general presence, indicated a radical derived from a molecule or an enzyme present in a significant quantity in the medium. The authors carried out experiments on different molecules such as cellulose, pectin, hemi-cellulose, or lignin, in a pure state, in order to compare their ESR spectra with those of irradiated fruits. The signal that was used to prove whether the strawberries had been irradiated or not belongs to a triplet radical induced in cellulosic parts of fruit. Consequently, it can be observed in all the fruit, more specially in their achenes, pips, or stones [11]. The left line (lower field) can be used as an identification test for irradiation.

Fresh fruit contains 10–20% of dry mass, while carbohydrates, mostly sugars, make 5–18% of this mass. In fresh fruit even of the highest sweetness, sugars never appear in the crystalline form. Crystalline domains appear in them only by drying in open air or freeze-drying. Natural or

Table 4.2 Content of sugars in fresh fruit

Dried fruit	Total sugars	Sucrose[a]	Fructose[a]	Glucose[a]	Fructose/ glucose	References
Pineapple	19.8	0.6	2.1	2.9	0.72	[12,13]
Banana	12.2	0.4	2.7	4.2	0.64	[14]
Papaya	9.9	1.6	3.1	5.1	0.61	[14]
Fig	47.9	0.07	24.4	26.9	0.91	[13,14]
Californian plum	12.4	1.2	1.4	2.7	0.52	[12,14]

[a]In grams per 100 grams of fresh fruit.

artificial dehydration of fruit increases the content of dry mass in fruit efficiently and is accompanied by an increase in carbohydrate concentration of up to 60–75% of the primary mass. The dehydration process stimulates spontaneous crystallization of sugars in the form of small crystalline domains composed mainly of glucose and fructose, and in some fruits like berries of mannose and sorbose. Table 4.2 illustrates the content of glucose and fructose in dried fruit taken as a model system. The content of glucose is distinctly higher than that of fructose, almost double in Californian plums; however, it is highest for figs (the last column in Table 4.2) [12].

For most fruit the contribution of sucrose to the overall sugar content is several times smaller than fructose, or glucose. The same situation occurs with dried fruit containing mannose or sorbose instead of glucose and fructose, as in most commercially available dried fruit. The dried fruit available in the market with crystalline domains of sugars (glucose, fructose, etc.) inside contains a lot of water which absorbs microwaves and makes ESR measurement very difficult or almost impossible. Fruit samples should therefore be further dehydrated before the experiment. Usually the samples of fruit are kept in a dryer at 35°C for up to 24 h, as recommended by CEN European standards EN 13708 [2], and EN 1787 [3,15].

It has been observed that some nonirradiated dried fruit show a weak single line in ESR, or show no ESR signal at all [16]. Fig. 4.4 shows the ESR spectrum recorded with dried fig, papaya, and banana. Nonirradiated dried papaya does not show any natural signal, while a weak ESR single line near $g_c(CF) = 2.004$ is observed in the case of nonirradiated fig and banana. Weak ESR signals with similar characteristics were observed in other nonirradiated fruits. The source of natural ESR signals in dried fruit is presumably due to paramagnetic impurities of

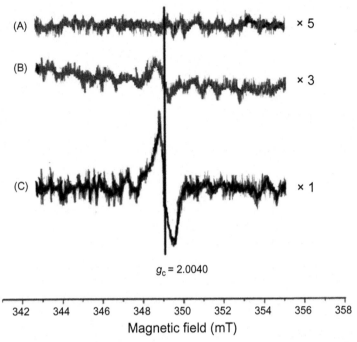

Figure 4.4 The ESR signals of dried fruits not irradiated: (A) papaya, (B) fig, and (C) banana. The vertical line denotes the center of ESR signals. Relative attenuations are shown on the right side.

dried fruit, stable free radicals, derivatives of semiquinones [17], or lignin [3], the constituents of fruit.

Dried fruit exposed to ionizing radiation exhibit comparatively intense complex ESR spectra. The intensity and substructure of multiline ESR spectra recorded after radiation treatment are very specific. Fig. 4.5 compares the ESR spectra of nonirradiated and radiation-treated dried pineapple [15].

Fig. 4.6 compares the ESR spectra of nonirradiated and irradiated dried banana [18].

Fig. 4.7 compares the ESR spectra of nonirradiated and irradiated dried fig [19].

Fig. 4.8 compares the ESR spectra of nonirradiated and irradiated dried papaya [15].

The intensity of the ESR signal of irradiated dried fig is clearly lower than those of dried banana, pineapple, and papaya. However, the signals are only a few times higher than average noise level, but are still sufficient to identify the signal as derived from irradiated material.

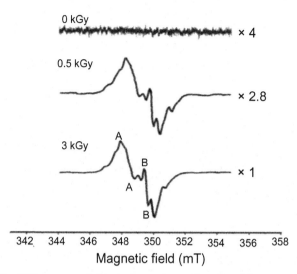

Figure 4.5 The ESR spectra of dried pineapple not irradiated (upper spectrum) and irradiated with a dose of 0.5 and 3 kGy respectively. Relative attenuations are shown on the right side. Nonirradiated dried pineapple does not show any signal, while irradiated pineapple shows two strong lines, A and B, overlapping the multiplet of smaller intensity. Even with a dose of 0.5 kGy the ESR signal is sufficiently intense, making the identification of radiation treatment possible. *Reproduced from G.P. Guzik, W. Stachowicz, J. Michalik, Nukleonika 60 (2015) 627–631 (Open access).*

Computer simulation studies may be used to relate the ESR spectra with the fructose and glucose content of the fruit. The ESR spectra obtained by computer superposition of commercial fructose and glucose taken in the proportion of 1:2.5 resembles the ESR spectra of irradiated pineapple, banana, and papaya (Fig. 4.9).

Fig. 4.10 shows the superposition of the ESR spectra of fructose and glucose taken in a proportion of 1:1. It resembles that of irradiated fig, with a similar content of fructose and glucose [15].

The ESR spectra recorded with dried fruit irradiated with doses of 0.5 and 3 kGy thus allow rough estimation of the critical doses applied to preserve them. The ESR signal intensity of five times higher than the signal noise level (5:1), as an accepted criterion, can be used to properly classify irradiated fruit.

The broadness of the ESR signals for different dried fruit commonly found in food markets and irradiated with 3 kGy gamma radiation varies for different fruits. The broadness of the ESR signal may therefore be used to classify irradiated fruits [2,7–9].

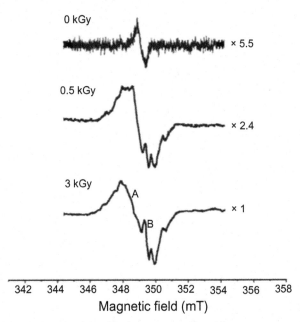

0 kGy

× 5.5

0.5 kGy

× 2.4

3 kGy

A

B

× 1

342 344 346 348 350 352 354 356 358

Magnetic field (mT)

Figure 4.6 The ESR spectra of dried banana not irradiated (upper spectrum), and irradiated with a dose of 0.5 and 3 kGy respectively. Relative attenuations are shown on the right side. Nonirradiated banana shows a weak natural ESR singlet with $g = 2.004$. The ESR spectra of irradiated samples of dried banana resemble that of dried pineapple, but with different intensities. This indicates the presence of similar radicals in both fruits.

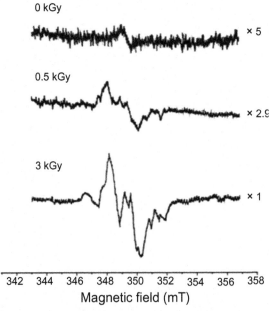

0 kGy

× 5

0.5 kGy

× 2.9

3 kGy

× 1

342 344 346 348 350 352 354 356 358

Magnetic field (mT)

Figure 4.7 The ESR spectra of dried fig not irradiated (upper spectrum) and irradiated with a dose of 0.5 and 3 kGy, respectively. Relative attenuations are shown on the right side. *Reproduced from G.P. Guzik, W. Stachowicz, J. Michalik, Nukleonika 60 (2015) 627–631 (Open access).*

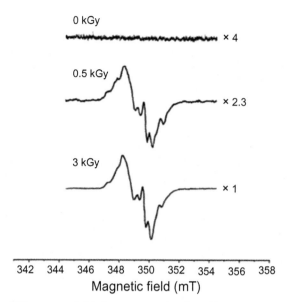

Figure 4.8 The ESR spectra of dried papaya not irradiated (upper spectrum) and irradiated with a dose of 0.5 and 3 kGy, respectively. Relative attenuations are shown on the right side. The ESR spectra of dried papaya are almost identical to those of dried pineapple, and are of about the same intensity, indicating identical fructose and glucose content in both fruits. *Reproduced with permission from Nukleonika.*

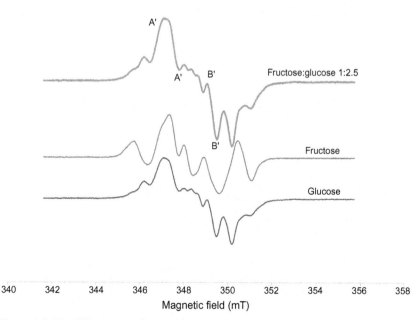

Figure 4.9 The ESR spectra obtained by computer superposition of commercial fructose and glucose (Aldrich reagents) taken in a proportion of 1:2.5 (upper spectrum) (irradiated with 3 kGy).

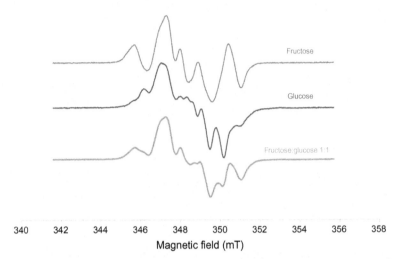

340   342   344   346   348   350   352   354   356   358
Magnetic field (mT)

Figure 4.10 The ESR spectra obtained by computer superposition of commercial fructose and glucose (Aldrich reagents) irradiated with 3 kGy and taken in the proportion of 1:1 (lower most). *Reproduced from G.P. Guzik, W. Stachowicz, J. Michalik, Nukleonika 60 (2015) 627–631 (Open access).*

ESR spectra obtained with crystalline sugars (a component of dried fruit) are used to propose the structures of dominating radiation-induced radicals [19–22]. There are publications reporting radicals appearing in spatial structures of D-fructose/D-glucose [23–25].

The exception among dried fruit exposed to radiation is California plum, which did not show any ESR signal even after irradiating with 3 kGy. The case is similar with dried peaches and some berries. Sugars contained in these fruits after drying do not appear in the form of crystalline domains. Therefore, they cannot contain any stable radicals which are only effectively trapped in the crystalline matrix. Consequently, specific ESR signals cannot be detected. This confirms the common belief that sugars which are definitely present in these fruit appear in the form of dense syrup only.

The dependence of ESR signal intensity on applied dose has been explored for pineapple, banana, papaya, and fig irradiated with different doses of gamma rays (0.5, 1, 2, and 3 kGys) and stored for 360 days at room temperature (Fig. 4.11) [15]. The intensity of the ESR signal has been found to increase linearly with increasing dose. This dependence is more pronounced in the case of papaya as compared to the three other

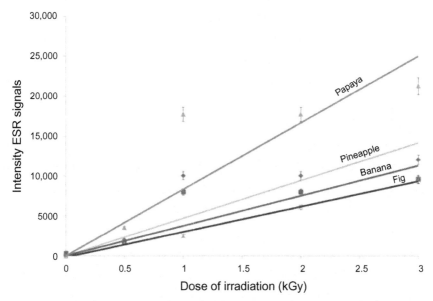

Figure 4.11 The dependence of ESR signal intensity with dose. Sample fruits were stored for 360 days after radiation treatment.

fruits studied. It reflects more effective trapping of radiation energy in the crystalline matrix, and indicates a different nature of radiation-induced radical centers in papaya. Tabner et al. [26] have observed this linear relationship in the case of irradiated grapefruit in the range 0–3 kGy. Helle et al. [27] have reported this linear dependence in the case of candied cherries, apples, pears, banana-chips, and the skin of orange and lemon.

The ESR signals of irradiated fruits have been found to decay with storage time. This time dependence is not governed by simple kinetic equations [28]. Fig. 4.12 shows the decay of the ESR signals in dried fruit irradiated with 1 kGy of gamma rays at ambient temperature with storage time [15]. The decay of radicals after a prolonged period of storage is found to be extremely slow. The high stability of radiation-induced radicals in dried fruit leads to ESR detection of radiation treatment on these commodities even after being stored for several years.

4.3 SPICES

Spices are mainly dried in the open air, and hence may be contaminated. This contamination may be due to many reasons. Bacteria, fungi, and insects

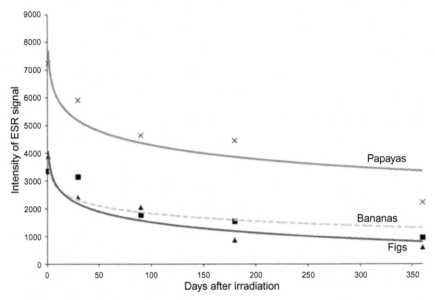

Figure. 4.12 Decay of the ESR signals in dried fruit irradiated with 1 kGy as a function of storage time up to 360 days. The relatively fast decay of radicals up to 50 days starting from the day of irradiation is clearly seen. The greatest decrease in the intensity is observed for the banana (43.0 ± 1%), and the lowest is for papaya (18.0 ± 1 %). *Reproduced from B.J. Tabner, V.A. Tabner, An electron spin resonance study of gamma-irradiated grapes, Radiat. Phys. Chem. 38 (1991) 523–531 (Open access).*

are some of the prime culprits in the contamination of spices. Treatment with ionizing radiation has been established to be more effective against bacteria than thermal treatment, as this does not lead to further contamination of the food item through chemicals [29,30]. Ahn et al. [31] have studied the effect of various sample pre-treatments on the ESR spectra of spices (turmeric, oregano, and cinnamon). Spectra acquired before and after irradiation were compared. They found that ESR analysis of irradiated food samples incorporating the alcoholic extraction process resulted in an improvement in the identification of irradiation status for irradiated spice samples relative to samples pre-treated with conventional heating and freeze-drying protocols. This is simply because the alcoholic extraction method gave rise to enhanced ESR signal intensity, resulting in clear detection of irradiation.

4.4 HERBS

ESR-based identification of radiation-induced radicals in herbs has been reported in the literature. Ginseng, which is a medicinal herb, is available

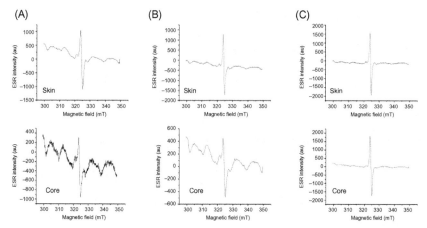

Figure. 4.13 The ESR spectra of 7 kGy irradiated fresh ginseng using different sample drying pre-treatments. (A) Freeze-drying, (B) alcoholic extraction, and (C) water washing and alcoholic extraction. *Reproduced from J.J. Ahn, K. Akram, D. Jo, J.H. Kwon, Investigation of different factors affecting the electron spin resonance-based characterization of gamma-irradiated fresh,white, and red ginseng, J. Ginseng Res. 36 (2012) 308–313 (Open access).*

in different forms. Ginseng and ginseng products have a natural potential risk of microbial and insect attack, and require an effective sterilization technique with the least compromise on quality attributes [32,33]. Ahn et al. [34] have reported ESR-based identification of radiation-induced radicals attempted using different sample pre-treatments (Fig. 4.13). They have discussed the possibilities of developing better ESR-based identification methods for different ginseng products. The effect of time of harvesting on radiation-induced ESR signals was also investigated, and there was a general increasing trend in ESR intensity of harvested irradiated samples aged at 4, 5, and 6 years (Fig. 4.14).

Slave et al. [35] have reported an ESR study of gamma-irradiated aromatic herbs dill and parsley.

Gamma irradiation could induce stable radicals in both samples, and they could be observed even after 50 days of irradiation. Further, they found that double integration of ESR spectra, corresponding to the area below the absorption curve, shows a monotonous increase with irradiation dose for both herbs. The two herbs, however, yield a different nature of increase, suggesting that the reaction mechanisms during irradiation are different.

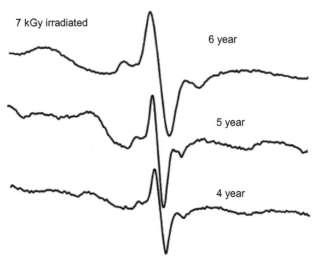

7 kGy irradiated

6 year

5 year

4 year

Figure. 4.14 Electron spin resonance spectral characteristics of 7 kGy-irradiated fresh ginseng harvested at different plant ages. *Reproduced from J.J. Ahn, K. Akram, D. Jo, J.H. Kwon, Investigation of different factors affecting the electron spin resonance-based characterization of gamma-irradiated fresh,white, and red ginseng, J. Ginseng Res. 36 (2012) 308–313 (Open access).*

4.5 BEVERAGES

ESR spectroscopy is suitably applied for free radical evaluation and to establish the antioxidant capacities of beverages. An evaluation of the antioxidant capacity of tea drinks has been recently reported by Popa et al. [36]. It is based on the measurement of changes of the ESR spectrum of radicals as a result of their interaction with antioxidants. The ESR measurements were carried out on five types of organic tea cultivated in Romania: *Mentha piperita* (MP), *Hypericum perforatum* (HP), *Achillea millefolium* (AM), *Rhamnus frangula* (RF), and *Calendula officinalis* (CO). Mn^{2+} in different environments was found in all dry tea leaves (Fig. 4.15). A supplementary ESR signal attributed to Cu^{2+} was observed in MP, HP, and AM tea. A weak contribution of Fe^{3+} was also confirmed for black tea and HP tea. Dry tea leaves show additional sharp line attributed to a semiquinone radical which disappeared in tea drinks (Figs. 4.15 and 4.16).

A scavenging effect which is an important mechanism of antioxidant activity in tea drinks was evaluated by ESR using stable free radical DPPH. The DPPH solutions were mixed with prepared tea drinks and the decline of ESR signal was monitored. The MP tea exhibited the

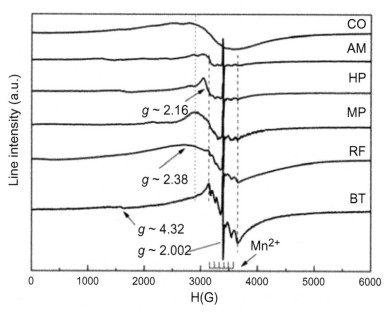

Figure. 4.15 *X*-band ESR spectra of dry tea leaves at room temperature. *Reproduced with permission from A. Popa, O. Raita, D. Toloman, Rom. J. Phys. 60 (2015) 1501–1507.*

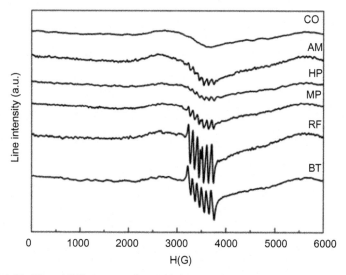

Figure. 4.16 *X*-band ESR spectra of tea drinks at room temperature. *Reproduced with permission from A. Popa, O. Raita, D. Toloman, Rom. J. Phys. 60 (2015) 1501–1507.*

highest antioxidant capacity comparable with black tea. Also, HP tea has an important antioxidant capacity. The weakest antioxidant capacity was observed for RF tea.

Free radicals in coffee beans which may be related to toxic effects can be examined using ESR spectroscopy. Free radical concentration and the effect of roasting on coffee beans of different origins have been recently reported by Krakowian et al. [37]. Higher free radical concentrations were reported for both the green and roasted coffee beans from Peru, compared with those originating from Ethiopia, Brazil, India, and Colombia.

4.6 CONCLUSION

ESR spectral features have been thus found to yield useful information on processed fruit, spices, and herbs. Quality criteria such as the antioxidant capacity of different beverages can be analyzed using ESR spectroscopy. The ESR analysis of irradiation dose of foodstuffs indicates the importance of this technique, with special reference to health effects. Further application of ESR spectroscopy to a broad range of food materials is expected to lead to societal benefits.

ACKNOWLEDGMENT

The authors wish to thank Prof. Jacek Michalik and Dr. Waclaw Stachowicz for their valuable suggestions.

REFERENCES

[1] EN 1786, Foodstuffs – Detection of Irradiated Food Containing Bone – Method by ESR Spectroscopy, European Committee for Standardisation (CEN), Brussels, 1996.
[2] EN 13708, Foodstuffs – Detection of Irradiated Food Containing Crystalline Sugar by ESR Spectroscopy, European Committee for Standardisation (CEN), Brussels, 2001.
[3] EN 1787, Foodstuffs – Detection of Irradiated Food Containing Cellulose by ESR Spectroscopy, European Committee for Standardisation (CEN), Brussels, 2000.
[4] J.A.P. Boshard, D.E. Holmes, L.H. Piette, An inherent dosimeter for irradiated food: papayas, Appl. Radiat. Isot. 22 (1971) 316–318.
[5] N.J.F. Dodd, A.J. Swallow, F.J. Ley, Use of ESR to identify irradiated food, Radiat. Phys. Chem. 26 (1985) 451–453.
[6] J.J. Raffi, J.-P.L. Agnel, L.A. Buscarlet, C.C. Martin, Electron spin resonance identification of irradiated strawberries, J. Chem. Soc. Farad. Trans. 1 (84) (1988) 3359–3362.
[7] J. Raffi, J.-P. Agnel, Electron spin resonance identification of irradiated fruits, Radiat. Phys. Chem. 34 (1989) 891–894.
[8] J. Raffi, J.-P. Agnel, Electron resonance identification of ionized foodstuffs, in: B. Catoire (Ed.), Electron Spin Resonance (ESR) Applications in Organic and Bioorganic Materials, Springer, Berlin Heidelberg, 1992, pp. 135–143.

[9] J. Raffi, J.-P. Angel, S.H. Ahmed, Electron spin resonance identification of irradiated dates, Food Tec 3/4 (1991) 26–31.

[10] J. Raffi, M. Kent, Methods of identification of irradiated foodstuffs, in: L.M.L. Nollet (Ed.), Handbook of Food Analysis Vol. 2, Residues and Other Food Component Analysis, Food Science and Technology Series 77, Marcel Dekker, Inc., New York, Basel, Hong Kong, 1996, pp. 1889–1906.

[11] J.J. Raffi, J.-P.L. Agnel, L.A. Buscarlet, C.C. Martin, J. Chem. Soc. Farad. Trans. I 84 (1988) 3359.

[12] A. Deken, R.D. Lindeman, The US Pharmacopeia Guide to Vitamins and Minerals, The United States Pharmacopeia Convention, Inc., Rockville, 1996.

[13] X. Shapiro, M. Ordman, E. Pivonka, Nutritionist V Database, Annual Report 2013, Produce for Better Health Foundation (051-608). Lancaster Pike 2013.

[14] US Department of Agriculture, Agricultural Research Service, Nutrient Data Laboratory. USDA National Nutrient Database for Standard Reference, Release 28. Version Current: September 2015, slightly revised May 2016, Internet: <http://www.ars.usda.gov/nea/bhnrc/ndl>.

[15] G.P. Guzik, W. Stachowicz, J. Michalik, Nukleonika 60 (2015) 627–631.

[16] B.J. Tabner, V.A. Tabner, An electron spin resonance study of gamma-irradiated grapes, Radiat. Phys. Chem. 38 (1991) 523–531.

[17] H.N. Swartz, J.R. Bolton, D.C. Borg, (Eds.), Biological Applications of Electron Spin Resonance, Wiley-Interscience, New York, 1972.

[18] Guzik, et al., Unpublished work, 2015.

[19] G. Vanhaelewyn, B. Jansen, E. Pauwels, E. Sagstuen, M. Waroquier, F. Callens, Experimental and theoretical electron magnetic resonance study on radiation-induced radicals in L-sorbose single crystals, J. Phys. Chem. A 108 (2004) 3308–3314.

[20] G. Vanhaelewyn, P. Lahorte, F. Proft, W. Mondelaers, P. Geerlings, F. Callens, Electron magnetic resonance study of stable radicals in irradiated D-fructose single crystals, J. Phys. Chem. Chem. Phys. 3 (2001) 1709–1735.

[21] G. Vanhaelewyn, B. Jansen, F.J. Callens, E. Sagstuen, ENDOR-assisted study of the stable EPR spectrum of X-irradiated α-L-Sorbose single crystals: MLCFA and simulation decomposition analyses, Radiat. Res. 162 (2004) 96–104.

[22] G. Vanhaelewyn, P. Lahorte, F. Proft, W. Mondelaers, P. Geerlings, F. Callens, Electron magnetic resonance study of stable radicals in irradiated D-fructose single crystals, Phys. Chem. Chem. Phys. 3 (2001) 1729–1735.

[23] G.P. Guzik, W. Stachowicz, J. Michalik, Study on stable radicals produced by ionizing radiation in dried fruits and related sugars by electron paramagnetic resonance spectrometry and photostimulated luminescence method – I. D-fructose, Nukleonika 53 (2008) S89–S94.

[24] G.P. Guzik, W. Stachowicz, J. Michalik, Study on organic radicals giving rise to multicomponent EPR spectra in dried fruits exposed to ionizing radiation, Curr. Top. Biophys. 33 (Suppl. A) (2010) 81–85.

[25] G.P. Guzik, W. Stachowicz, J. Michalik, EPR study on sugar radicals utilized for detection of radiation treatment of food, Nukleonika 57 (2012) 545–549.

[26] B.J. Tabner, V.A. Tabner, An electron spin resonance study of gamma-irradiated grapes, Radiat. Phys. Chem. 38 (1991) 523–531.

[27] N. Helle, B. Linke, G.A. Schroeiber, K.W. Bogl, Nachweis der gamma Bestrahlung von Trockenfrüchten, Bundesgesundheitsblatt 35 (1992) 179–184.

[28] P.B. Ayscough, Electron Spin Resonance in Chemistry, Methuen & Co Ltd (1967),

[29] T.B. Tjaberg, B. Underdal, G. Lunde, The effect of ionizing radiation on the microbial content and volatile constituents of spices, J. Appl. Bacteriol. 35 (1972) 473–478.

[30] P. Loaharanu, Status and prospects of food irradiation, Food Technol. 52 (1994) 124–131.

[31] J.J. Ahn, B. Sanyal, K. Akram, J.-H. Kwon, Agric. Food Chem. 62 (2014) 11089–11098.
[32] J.H. Kwon, J.M. Belanger, J.R. Pare, Effects of ionizing energy treatment on the quality of ginseng products, Radiat. Phys. Chem. 34 (1989) 963–967.
[33] J.H. Kwon, J. Lee, C. Waje, J.J. Ahn, G.R. Kim, H.W. Chung, et al., The quality of irradiated red ginseng powder following transport from Korea to the United States, Radiat. Phys. Chem. 78 (2009) 643–646.
[34] J.J. Ahn, K. Akram, D. Jo, J.H. Kwon, Investigation of different factors affecting the electron spin resonance-based characterization of gamma-irradiated fresh, white, and red ginseng, J. Ginseng Res. 36 (2012) 308–313.
[35] R.M. Slave, C.D. Negut, V.V. Grecu, ESR on some gamma-irradiated aromatic herbs, Rom. J. Phys. 59 (2014) 826–833.
[36] A. Popa, O. Raita, D. Toloman, Rom. J. Phys. 60 (2015) 1501–1507.
[37] D. Krakowian, D. Skiba, A. Kudelski, B. Pilawa, P. Ramos, J. Adamczyk, et al., Application of EPR spectroscopy to the examination of pro-oxidant activity of coffee, Food Chem. 151 (2014) 110–119.

CHAPTER 5

ESR Spectroscopy for the Identification of Irradiated Fruits and Vegetables

K. Akram, U. Farooq and A. Shafi
University of Sargodha, Sargodha, Pakistan

Contents

5.1 INTRODUCTION

Food safety is considered to be one of the major emerging demands of the current era, due to the awareness of diet and various established health issues. Among all technologies ensuring the provision of safe food to consumers, food irradiation has achieved a promising novel status all over the world. Food irradiation technology involves the careful exposure of food (either packaged or in bulk) to ionizing radiation for a certain time period with controlled dosing to achieve the desired goals of food safety or preservation. Consumers are well aware of the effectiveness of food irradiation, and its potential to reduce the risk of food-borne diseases and increase the shelf life of food commodities. During storage of bulbs, tubers, and nuts, retardation of sprouting and rooting can be achieved at low doses of irradiation ($<1\,kGy$). For the development of sterilized food for special purposes, high doses of irradiation (>10 kGy) may prove more effective. Over a wide range of applications, irradiation in the range of 1–10 kGy, alone or in combination with other potential technologies, can reduce microbial load, kill insect eggs and larva, affect enzyme activity, and improve the sensory properties of food commodities [1].

Among foods, fruit and vegetables are indispensable for the people of all ages, because these provide bioactive compounds in addition to the provision of basic nutrition. Due to the highly perishable nature of fruit and vegetables, these are more vulnerable to attack by microorganisms which result in a shorter shelf life and make food unsafe for consumption. Hence, in order to prolong the keeping quality and to ensure food safety, irradiation (with electron-beam, X-rays, and gamma-rays) can be used as an effective preservative technique [2].

5.2 NEED TO REGULATE FOOD IRRADIATION

The first industrial application of irradiation technology for food on a commercial scale was started in Stuttgart, Germany, in 1957 by Gewurzmuller and Co. [3]. But after 1959, the irradiation of food was prohibited by the food laws. The prohibition was not because of any food safety issue, but only because more research-based confirmation or validation on safety and wholesomeness was required for this new preservative technique [4]. Then the major international organizations, such as FAO, IAEA, and WHO formed a joint FAO/IAEA/WHO expert committee in 1961 to discuss the necessary legal and safety aspects of irradiated foods. A special assignment was also given to Professor J.F. Diehl for the development of methods to identify radiation-induced markers. So, in this way, the identification of food irradiation was started from this institution [5,6].

Currently more than 55 countries have regulated the use of radiation technology for different food materials, and irradiated food products are available in the commercial markets of more than 30 countries [1]. Although international health authorities have investigated and confirmed the technical efficacy of food irradiation for the health and safety of consumers [7], for the safe use and distribution of irradiated foods, and for consumer awareness, there is a mandatory requirement for labeling the treated foods or food containing irradiated ingredients as "irradiated" or "treated with ionizing radiation," with the elective use of the "Radura" mark (Fig. 5.1) [8]. Irradiation also serves as a quarantine treatment for the long-term preservation of food commodities. For export purposes,

Figure 5.1 "Radura" mark for the labeling of irradiated food.

there is also the need for an identification method to satisfy the different well-defined strict regulations related to trade in irradiated food products in some advanced countries [9]. In the absence of a reliable identification method, the acceptance of the marketed food (irradiated) and the validation of quarantine requirements are difficult [10,11].

The most important aspect of a detection method is its specificity against well-characterized radiolytic or radiation-induced markers. The concentration of these radiation-induced detection markers should be dependent on the applied radiation dose, and there should be a minimal or no effect of post-irradiation storage conditions and different food processing treatments. The situation becomes more difficult as the concentration of radiation-induced detection markers mainly depends upon the availability and stability of radical species, which is affected by several parameters including oxygen and moisture content, temperature during the application of radiation, and composition of the food matrix. The European Committee of Standardization (CEN) has standardized various methods to characterize the irradiation status of foods, and the same methods are also endorsed by the Codex Alimentarius. On the basis of practical application, these methods are further categorized into screening and confirmatory tests [12].

Carbohydrates are one of the major compositional components of plant-based food commodities. The radiolysis of carbohydrates generates acids and carbonyl groups as a result of complex changes. The viscosity of foods is modified as a result of this radiolysis, due to changes in molecular weight or degree of methylation [13]. However, the results for reliable identification of irradiated foods by analyzing the breakdown products of carbohydrates are not satisfactory, due to the fact that the products under observation are not radiation specific [14]. Moreover, it had been observed that factors related to origin, type, variety, ripening stage, and conditions related to processing and storage environment had a significant role in changes in viscosity and concentration of the radiolytic products, as compared to the changes due to irradiation only [12].

Among other available methods (physical, chemical, and biological) for the reliable identification of irradiated foods, electron spin resonance (ESR) spectroscopy of radiation-induced radicals is a promising physical technique to characterize the irradiation status of various food materials.

5.3 ELECTRON SPIN RESONANCE SPECTROSCOPY

ESR spectroscopy is a spectroscopic technique that is specified by the European Standards for the identification of paramagnetic centers (radicals) of irradiated foods. The ESR spectroscopy detects the free radicals (e.g., paramagnetic centers) generated by the irradiation treatment of food material. In this technique, a significant difference between the energy intensities of the electron spins $ms = \pm\frac{1}{2}$ is created through subjection to an external magnetic field. Resonance produced as a result of applied microwave energy is absorbed, and the absorbance is measured. The ESR spectrum is presented as a derivative of microwave absorption in relation to the value of the magnetic field. The frequency of the external magnetic field is measured as the g value of the ESR signal. The ESR signal g value can be calculated as $h\nu/\beta B_0$, where h is Planck's constant, ν is the microwave frequency, β is the Bohr magneton, and B_0 is the magnetic field. This g value of ESR signal is used for detection purposes.

ESR spectroscopy was first applied by Gordy et al. [15] to study the generation of radicals in biological matrices. However, practical application of the ESR-based technique to identify irradiated food was reported after 16 years [16]. This is only possible when there is limited or low reactivity of the radicals with each other, as well as with water, and such a situation can be found in the solid and dry components of the food [17–19]. ESR is an easy and convenient technique, due to the fact that the method is very simple, and there is no need for sample preparation. Recently, there is an increasing tendency towards the adaptation of EPR methodology for the detection of various types of irradiated foods. The dried components of the sample matrix are mostly used for the identification, as radicals have limited activity because of the rigid structure. The free radicals, after induction into the food matrix on irradiation treatment, are usually very reactive and least stable especially in an environment of high moisture, where they react with other reactive species and food components. This may lead to different chemical reactions and, as a result, various radiolytic products may be formed. The radiation-induced markers used to identify irradiated food materials should have a stability which is comparable to the storage life of the food [12].

Different food processing techniques applied in the food industries produce various types of radicals; however, there are three types of radiation-specific free radicals that can produce specific ESR spectra as proof of irradiation [20,21].

- *Hydroxyapatite radicals* from bone-containing foods (animal origin). CO_2^{-1}, CO_3^{-3}, and CO_3^{-1} are the main radiation-induced radicals in the carbonate matrix.
- *Cellulose (cellulosic) radicals* in food materials of plant origin contain cellulose such as seeds, peels, and some herbs and spices.
- *Sugar radicals* from dried fruit and dried plant materials contain different sugars in crystalline form.

Considering these three radiation-induced radicals, the European Union has standardized three ESR methods primarily targeting food containing bone (EN 1786: 1997), cellulose [21], and crystalline sugar [22]. These European standards are also endorsed by the Codex Alimentarius as General Codex Methods for the detection of irradiated foods. The results of identification are usually obtained in the form of qualitative results and through this qualitative discrimination, irradiated and nonirradiated samples can be separated. However, the irradiation dose can only be estimated by using re-irradiation techniques with least variations in the sample composition [23]. The scope of this chapter is limited to the ESR spectra from radiation-induced cellulosic and crystalline sugar radicals. The presence of these signals is considered as conclusive evidence of irradiation history; however, their absence may be due to issues with the stability and availability of cellulosic and crystalline sugar radicals [12].

5.4 DETECTION OF IRRADIATED FRUITS AND VEGETABLES USING ELECTRON SPIN RESONANCE TECHNIQUES

The irradiation of foods of plant origin generates free radicals in cellulose and crystalline sugars, and the unpaired electrons in these radicals become the source for detection of irradiation of fruit and vegetables [23]. However, the stability of these radiation-induced free radicals is an important factor for the effective use of these radical species as radiation-induced markers. The major reasons for the instability of free radicals include the high moisture contents of fresh fruit and vegetables, and the noncrystalline nature of sugar present in fruit and vegetables [2]. Clear ESR spectra can be obtained by selecting the hard parts of irradiated fruits and vegetables like skin, seeds, and shell for detection purposes [24]. The sample for ESR analysis is prepared by removing moisture, either by freeze-drying or other techniques, as the specificity of ESR signals is affected by high moisture contents [21]. The free radicals disappear in high moisture contents very rapidly, and can react easily with other components present in moisture, as compared to solid or semi-solid food materials [25,26].

In different fresh agricultural commodities, ESR spectra may give a silent signal (Fig. 5.2) for a nonirradiated sample. Nonirradiated

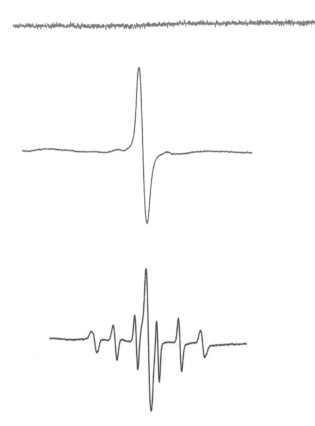

Figure 5.2 ESR spectra in nonirradiated foods of plant origin (Above: ESR silent (no signal); Middle: natural signal usually in food of plant origin; below: effect of Mn^{2+}).

samples may also provide a central ESR signal (Fig. 5.2) that was attributed by various scientists to the generation of semiquinone radicals due to the oxidation of polyphenolic components from the plant matrix [27]. This main ESR spectral signal increases with an increasing dose of irradiation; however, the signal is also found to be sensitive to other food processing techniques (e.g., drying processes) [28]. By the exposure to irradiation, the intensity of the central signal is enhanced, with the development of two side peaks that are usually attributed to radiation-induced cellulose radicals (Fig. 5.3). The mutual distance (H2 − H1 ≈ 6 mT) of side peaks (g = 2.020 and g = 1.985) of the ESR signal depends upon the radiation treatment, but lack of this specific signal does not mean that the sample is nonirradiated [21]. It is

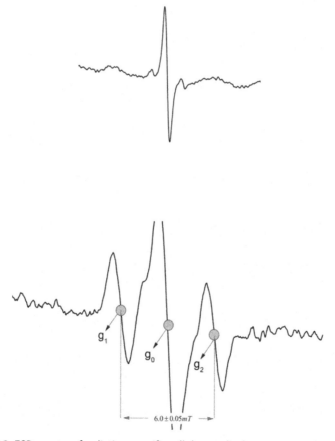

Figure 5.3 ESR spectra of radiation specific cellulose radical.

estimated that the left side peak gives the indication of cellulose radicals, whereas the right side peak signal gives the indication of lignin radicals [28]. Various scientists observed radiation-induced cellulosic radical signals in paprika powder [29], dried parts of different fruit such as achenes, pips, shells, stalks, or seeds [30], citrus fruit skins, skin components and stalks [31,32], and fruit cell walls as well [33–35]. The presence of transition metals (Fe^{3+} and Mn^{2+}) in plant matrices also affect the specificity of the ESR signal [29]. A Mn^{2+}-induced sextet signal (Fig. 5.2) may also arise from the nonirradiated food matrices that increases on irradiation and makes detection difficult, and in some case impossible [12,29]. Korkmaz and Polat [36] experienced difficulty in characterizing the irradiation status of dry broad bean samples. On irradiation there was a spectrum showing an equally spaced sextet from Mn^{2+} ions and a single resonance line from radiation-induced radicals. The intensity of the free radical signal was dependent on the applied irradiation dose.

Kwon et al. [23] studied the radiation-induced markers in dried cabbage, carrot, chunggyungchae, garlic, onion, and green onion using ESR spectroscopic techniques. Radiation specific cellulosic radicals were detectable by ESR spectra even after six months of storage. In addition to the main ESR signal that was also found in nonirradiated samples, radiation treatment generated two side peaks (from cellulose radicals). Jo et al. [31,34] noted the successful application on the basis of radiation-induced cellulosic radicals in oranges, grapefruits, mandarins, limes, and pineapple after freeze-drying of the samples. The specification used for the determination of ESR signals in the European ESR protocol were also found not to be satisfactory by some researchers [37], and they observed that saturation behavior of the ESR signal is highly dependent upon different factors including moisture, fineness of the sample powder (if any), and the filling factor of the sample [38]. This observation clearly demonstrates that the fixed microwave power value, as given in the protocols recommended by CEN (0.4–0.8 mW), cannot be assumed, instead this value must be determined from sample to sample.

The effective determination of a radiation-specific cellulosic ESR signal depends upon various factors, such as sample physicochemical parameters like moisture content and temperature during irradiation treatment, irradiation dose, and storage conditions of the sample [39]. Food samples with low moisture contents do not require any pre-treatment to remove moisture for ESR-based characterization of irradiation history [40,41].

However, food materials with high moisture contents (e.g., fresh fruit and vegetables) require some effective sample pre-treatments to reduce moisture levels without effecting the radiation-induced cellulosic radicals. The available choices for moisture lowering pre-treatments include freeze-drying [42], alcoholic extraction [33,35], oven-drying [24], or other techniques [43]. For irradiated fresh mangoes, Kikuchi et al. [44] have reported the detection of radiation-specific ESR after one week of storage by using the fine powder of lyophilized mango pulp. Kikuchi et al. [43] described another effective technique for identification of irradiated fresh papaya through ESR spectra at a liquid nitrogen temperature of 77 K. Among all these techniques to address the problem of high moisture contents, solvent extraction, especially alcoholic extraction, may prove to be effective and convenient techniques because of the short time taken and cost-effectiveness [42,45]. De Jesus et al. [33] reported an alcoholic extraction technique to reduce moisture contents with successful applications for fresh kiwifruit, papaya, and tomato. A similar technique was also used by Delincée and Soika [35] showing improved ESR signal from cellulose radicals in strawberry, papaya, and some spices. The practical applicability of an alcoholic extraction technique was also found satisfactory when compared with other moisture lowering techniques for different food samples [31,34,44–46]. Akram et al. [25] gave a method in which irradiated liquid sauce samples were washed with water before extraction with alcohol. The water washing, coupled with alcoholic extraction, increased the detection of cellulosic radical signals, and the results were better when compared with routine freeze-drying and alcoholic extraction pre-treatments. Ahn et al. [47] reported that water washing with alcoholic extraction pre-treatment is also effective in improving ESR signals of radiation-induced cellulosic radicals in turmeric, oregano, and cinnamon.

Crystalline sugar radicals are formed in dried fruit and vegetables upon irradiation, and these radicals give multicomponent ESR spectra (Fig. 5.4). Due to the difference in the sample composition, and mono- and disaccharide crystallinity, a variation in ESR signals (centered at $g = 2.003$) are obtained [22]. The intensity of radiation specific ESR signals depends on the applied radiation dose and is easily identified in food matrices of low moisture where enough concentration of sugar exists in crystalline form [48]. These crystalline sugar radicals induced by specific radiations are stable during a storage period of several months, and can give a clear ESR signal to characterize the irradiated dried fruit and vegetables [49]. Dried fruit such as figs, raisins, mangoes, and papayas having crystalline sugars could provide

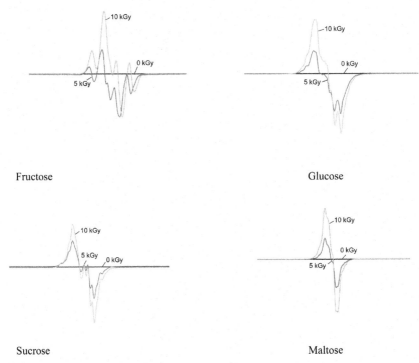

Figure 5.4 ESR spectra of radiation specific crystalline sugars.

radiation specific ESR signals, whereas the absence of radiation treatment relates to the absence of sugar radical signals. This confirms the specificity of these signals towards radiation treatment [50]. The ESR spectra only give specific signals in the case of specific compounds e.g., crystalline sugar, and the specificity of the ESR signal depends upon the type and crystallinity of the sugar [48]. In different fruit samples, the radiation-induced radicals of crystalline sugars usually give variable structures of ESR spectra, due to their complex and highly variable composition [51]. De Cooman et al. [52] identified successes in the structure of sucrose irradiated single crystals by using density functional theory and calculations of electron magnetic resonance parameters. One of the major stable radiation-induced radicals (radical T_2) has been detected as a glucose-centered radical with the major part of spin density at the C_1 position. The structure of the radical involves scission of the glycosidic bond linkage, and the presence of a carbonyl group at the C_2 position. A small spin density at C_1 and the enhanced g factor anisotropy are well-known characteristics of the radical achieved by the formation of a carbonyl group at the C_2 position. Another radical, T_3, was also explored which was most likely identical in chemical structure to T_2, but with a small difference in the orientation of neighboring hydroxy groups.

Da Costaa et al. [53] described the dosimetric characteristics of different sugars (sucrose and dextrose in powdered form) by measuring the free radical densities through ESR. It was observed that the ESR signal obtained from both irradiated and nonirradiated samples was the same at a grain size of $<80\,\mu m$, but the relative ESR absorption intensity increased as the powder size decreased. A typical powder spectrum obtained for sucrose samples after gamma irradiation consists of a single line centered at $g = 2.0014$. The spectrum of sucrose is complex, and the complex behavior is because of super hyperfine interactions of more than one radical produced in the structure of sucrose, and an average of all possible crystal orientations. Whereas glucose has a lot of hydroxyl groups, it has a small and flexible molecular structure so it gives a definite spectra. Comparison showed that sucrose is more sensitive than dextrose, but for both sugars the free radical ESR signal increases linearly with the radiation dose over a wide range $(0.1–100\,Gy)$. The minimum detectable dose is $0.1\,Gy$.

Vanhaelewyn et al. [54] tried to explore the nature of stable radicals in irradiated D-fructose single crystals. For this purpose, D-fructose powder was irradiated at ambient temperature in the dose range $1–20\,kGy$, and the corresponding electron paramagnetic resonance (EPR) spectrum series as a function of dose was subjected to MLCFA (maximum likelihood common factor analysis). After X-irradiation the single crystals have been investigated with electron nuclear double resonance (ENDOR) and ENDOR-induced EPR (EI–EPR) at $60\,K$ temperature. The EPR study of D-fructose single crystals in X-band revealed a weak g tensor anisotropy around $g = 2.005$. Two distinct absorption patterns were observed corresponding to two different radicals, and these different radicals are responsible for hyperfine interactions. The EI–EPR experiments evidently exposed two dominant spectrum components which interact with three hydrogen nuclei.

Malec-Czechowska et al. [49] tried to characterize the irradiation history of different mushroom samples using EPR spectroscopy. Both cellulose and crystalline sugar-like radicals were found and used for detection purposes. Different parts of the same mushrooms showed varying dose–response, where some mushroom samples were difficult to characterize using ESR techniques. Bayram and Delincée [55] studied the identification of irradiated Turkish foodstuffs using different physical detection methods and found that dried apricot (*Prunus armeniaca*), dried fig (*Ficus carica*), and raisins (*Vitis vinifera*) were easily identifiable for irradiation treatment on the basis of radiation-induced sugar radicals. Sanyal et al. [56], performed an investigation of irradiated rice samples and found an asymmetric ESR spectrum ($g = 2.005$) that was from radiation-induced radicals in rice starch. The intensity of ESR signals was found to be dose dependent; however, these signals were not stable during storage. The scientists further investigated the relaxation characteristics and thermal behavior of the radiation-induced radicals, and reported that the difference in microwave saturation behaviors of the radiation-induced signals ($g=2.004$) in irradiated and nonirradiated rice samples could provide conclusive evidence to identify radiation treatment even after the radiation specific ESR signals disappeared. Akram et al. [12] conducted a thorough ESR-based investigation on different parts of irradiated dried mushroom (*Lentinus edodes*). Radiation specific ESR signals were found at an irradiation dose of more than 2 kGy, where signals were clearer in mushroom cap samples. The scientists further observed that these radiation-induced ESR signals were sensitive to processing conditions involving high temperatures. Guzik et al. [57] observed complex sugar radical EPR spectra on irradiation of dried banana, pineapple, papaya, and fig, even 1 year after irradiation. Glucose and fructose radicals were the major paramagnetic products, and the similar ESR spectra were possible to reconstruct by superposition of the spectra of irradiated glucose and fructose. The intensity of radiation specific ESR signals showed a drastic decrease during the first month after irradiation. However, the signals were quite stable after prolonged storage, and were detectable even after 1–3 years of post-irradiation storage.

5.5 DOSE ESTIMATION

There are various factors related to food composition and the environment in which food is processed and stored, which affect the overall detectability of radiation specific radicals. The difficult control of these factors makes it impossible to properly quantify the exact concentration of radiation-induced radicals and relate them to the applied irradiation dose. This means only qualitative discrimination of irradiated and nonirradiated foodstuffs is possible [12]. However, different scientists tried to establish the relation between the intensity of the analyzed signal and initial applied dose. Ghelawi et al. [58] studied dose dependent radiation-induced cellulosic radicals in irradiated dates, where the signals were stable up to 27 months during storage. All 21 samples were easily detectable in a blind trial, and dose was estimated with an error up to 0.5 kGy.

Slave et al. [59] identified the relationship of concentration versus irradiation dose and a saturated exponential dependence by ESR spectroscopy while using gamma irradiated dill and parsley (two very used aromatic herbs). Target doses applied to the samples were 1, 3, 5, 7, 9, and 15 kGy, respectively. The ESR spectra showed the double integrated area below the absorption curve which indicated a monotonous increase with irradiation dose for both herbs. The double integrated area of the ESR signal is also proportional to the number of radicals in the sample. This behavior of spectra showed the dose dependency, and its complex signal reflected the existence of at least one dominant radical type. The observed spectra showed stability over time. The gamma-irradiated induced stable radicals can be observed even after 50 days of irradiation. It was also observed that two plants have quite different behaviors, suggesting that the reaction mechanisms during irradiation are different. In contrast, Braşoveanu et al. [60] observed that free radical concentration increased with irradiation dose and decreased with storage period while investigating the levels of free radicals in irradiated wheat flour and wheat bran. Three types of cellulose-like typical signals were shown by the ESR spectra. The ESR technique in detecting electron beam irradiation is strongly dependent on the stability of the induced free radicals, and this stability remains for a limited period after irradiation that could be shorter than the shelf life of the products. Abdel-Fattah et al. [61] reported the successful discrimination of irradiated and nonirradiated cumin, and also estimated the absorbed dose. An additive re-irradiation method was used to check the dose— response function in cumin samples, and the initial dose was assessed by back-extrapolation. EPR signal/dose curves were fitted using third degree polynomial and exponential functions, where a third degree polynomial fit gave a satisfactory result without considering the decay of radiation-induced free radicals in the complex food matrix.

5.6 VALIDATION

The European Standard EN 13708: 2002 [22] is based on radiation-induced crystalline sugar radicals, and was validated through two inter-laboratory tests conducted by the Community Bureau of Reference (CBR) and the German Federal Institute for Health Protection and Veterinary Medicine (BgVV). The first inter-laboratory validation test was carried out on dried papayas and raisins (the irradiation dose ranged from 0.5 to 7 kGy, or nonirradiated). In this test 126 samples for each of dried papayas and raisins were used. The results of 21 laboratories under these conditions indicated that specific ESR signals appeared. From all the samples, 7 samples from raisins and 2 from dried papayas were identified as nonirradiated, whereas only 1 nonirradiated sample from raisins showed as irradiated by the ESR spectrum. All the remaining samples showed a specific ESR signal at a 0.5 kGy irradiation dose for irradiation identification [17,50]. The other validation study [62] for sugar radicals was performed on dried mangoes and figs (184 samples for each), either irradiated to about 1 to 5 kGy, or nonirradiated. This study was performed through 17 laboratories by the German Federal Institute for Health Protection and Veterinary Medicine (BgVV). Using similar experimental conditions, the results of these studies revealed only two irradiated samples of dried figs were identified as nonirradiated, whereas all other samples showed specific ESR spectra for the identification of sugar radicals.

The European Standard EN 1787: 2000 [21] is based on radiation-induced cellulose radicals and was validated by two inter-laboratory tests with pistachio nut shells [50,63], one inter-laboratory test with paprika powder [64,65], and one with fresh strawberries [64]. In the first trial of 84 samples, 15 were found to be false positives, while 2 were determined as false negatives. In a second trial of 68 samples of pistachio nut shells, only one sample was determined as a false positive. In an inter-laboratory test with 168 samples of paprika powder, only one was found as a false positive. In an inter-laboratory test with 184 samples of fresh strawberries, 7 were determined as false negatives, while 2 were inconclusive.

5.7 LIMITATIONS

The application of ESR spectroscopy is nondestructive, user-friendly, simple, convenient, and time efficient for dried food samples in which cellulose radicals and crystalline sugar radicals are produced on irradiation. However, in high moisture foods, sugar cannot exist in crystalline form, and cellulose radicals cannot be measured as the O—H dipole of the water molecule absorbs the microwave energy. In this case, the determination of cellulose radicals requires an effective sample drying technique without disturbing the concentration and specificity of the radiation-induced radical [25].

Foods of plant origin are highly variable in their composition, where the presence of radiation specific ESR signals (crystalline sugar and cellulose radicals) is considered as conclusive evidence of irradiation treatment; however, the absence of such signals does not guarantee the absence of irradiation treatment of food [12]. In the case of the absence of sugar in crystalline form, it is possible for irradiated dried fruit (e.g., apricot or prune) to be silent for any radiation specific signal, even at higher irradiation doses [66]. Similar problems were observed in the case of cellulose radicals in the determination of different fresh and dried vegetables (e.g., black tea, hazelnut) [55].

Mn^{2+} is naturally present in many foods of plant origin and can disturb the ESR-based identification of food containing cellulose and sugar signals. Ahn et al. [29] reported the effect of soy sauce powders rich in Mn^{2+} on the ESR-based determination of radiation-induced cellulosic radicals in red pepper powder samples and crystalline sugar radicals in pak choi samples. There was a clear decline in the intensity of the left and the central peaks of ESR spectra of cellulosic radicals with increasing concentrations of soy sauce powders, while the Mn^{2+} signal showed an increase in the right peak by overlapping the cellulose signal. However, the radiation-specific sugar signals showed the least effect of Mn^{2+}.

The stability of radiation-induced signals over storage time is also a crucial aspect, as many food items showed radiation specific signals from a few weeks to a few months. Various scientists [67,68] tried different approaches such as the study of thermal treatments and EPR saturation behavior to get clues to the irradiation history of samples.

Moreover, post-irradiation storage conditions and storage periods greatly influenced the specificity of ESR spectra. ESR-based identification is strictly dependent on free radical stability or post-irradiation storage duration, e.g., within three weeks of post-irradiation storage free radicals remain stable and give clear ESR spectra.

5.8 SUMMARY

Reliable detection of irradiated foods is required for the widespread application of food irradiation technology. However, all available methods have limitations, and there is not a single method with the potential to analyze all food materials with respect to their irradiation history. Food materials of plant origin containing cellulose or crystalline sugar could be effectively characterized using ESR spectroscopy. However, the complex nature of food matrices, high moisture contents, low concentration and stability of radiation-specific radicals, and their effective detection during storage under practical conditions are major constraints for the reliable identification of irradiated food using ESR spectroscopy. ESR spectra based on radiation-specific cellulosic or crystalline sugar radicals confirm the radiation treatment, but the lack of a specific ESR signal is not evidence of nonirradiation of a food.

ACKNOWLEDGMENT

We would like to thank Prof. Dr. Joong-Ho Kwon (Kyungpook National University, Korea) for his worthy guidance, encouragement, and advice.

REFERENCES

[1] J. Farkas, C. Mohácsi-Farkas, History and future of food irradiation, Trends Food Sci. Technol. 22 (2) (2011) 121–126.
[2] E.F.O. De Jesus, A.M. Rossi, R.T. Lopes, Identification and dose determination using ESR measurements in the flesh of irradiated vegetable products, Appl. Radiat. Isot. 52 (5) (2000) 1375–1383.
[3] K.F. Maurer, Zur Keimfreimachung von Gewürzen, Ernährungswirtschaft 5 (1958) 45–47.
[4] J.F. Diehl, Safety of Irradiated Foods, CRC Press (1999).
[5] D.A.E. Ehlermann, Eröffnung und Begrüßung. Lebensmittelbestrahlung – 5. Deutsche Tagung. Berichte der Bundesforschungsanstalt für Ernährung. BFE–R—99–01, Bundesforschungsanstalt für Ernährung, Karlsruhe (Deutschland) (1999) 1–4.
[6] IAEA, Food irradiation. In: Proceedings of the International Symposium on Food Irradiation jointly organized by the International Atomic Energy Agency and the Food and Agriculture Organization of the United Nations and held in Karlsruhe, IAEA, Vienna, Austria, 6–10 June 1966. pp. 956, 1966.
[7] C.H. McMurray, E. Stewart, R. Gray, J. Pearce, Detection Methods for Irradiated Foods: Current Status, Royal Society of Chemistry (1996).
[8] CODEX STAN 1-1985, Amend. 7-2010. Codex General Standard for the Labelling of Prepackaged Foods. Food and Agriculture Organization/World Health Organization, Codex Alimentarius Commission, Rome, Italy.
[9] E. Marchioni, Detection of irradiated foods, Food Irradiat. Res. Technol. (2006) 85–103.
[10] S.K. Chauhan, R. Kumar, S. Nadanasabapathy, A.S. Bawa, Detection methods for irradiated foods, Compr. Rev. Food Sci. Food Saf. 8 (1) (2009) 4–16.
[11] I.S. Arvanitoyannis, Irradiation of Food Commodities: Techniques, Applications, Detection, Legislation, Safety and Consumer Opinion, Academic Press (2010).
[12] K. Akram, J.J. Ahn, J.H. Kwon, Identification and characterization of gamma-irradiated dried lentinus edodes using ESR, SEM, and FTIR analyses, J. Food. Sci. 77 (6) (2012) C690–C696.

[13] J. Farkas, A. Koncz, M.M. Sharif, Identification of irradiated dry ingredients on the basis of starch damage, Int. J. Radiat. Appl. Instrum. Part C. Radiat. Phys. Chem. 35 (1) (1990) 324–328.

[14] H. Delincee, D.A.E. Ehlermann, Recent advances in the identification of irradiated food, Int. J. Radiat. Appl. Instrum. Part C. Radiat. Phys. Chem. 34 (6) (1989) 877–890.

[15] W. Gordy, W.B. Ard, H. Shields, Microwave spectroscopy of biological substances. I. Paramagnetic resonance in X-irradiated amino acids and proteins, Proc. Natl. Acad. Sci. U.S.A. 41 (11) (1955) 983–996.

[16] J.A.P. Boshard, D.E. Holmes, L.H. Piette, An inherent dosimeter for irradiated foods: papayas, Int. J. Appl. Radiat. Isot. 22 (5) (1971) 316–318.

[17] J. Raffi, Electron Spin Resonance Intercomparison Studies on Irradiated Foodstuffs, Commission of the European Communities(Report EUR/13630/EN), Luxembourg, 1992.

[18] J.J. Raffi, S.M. Benzaria, Identification of irradiated foods by electron spin resonance techniques, J. Radiat. Sterilizat. 1 (1993) 282–304.

[19] A.J. Swallow, Need and role of identification of irradiated food, Int. J. Radiat. Appl. Instrum. Part C. Radiat. Phys. Chem. 35 (1) (1990) 311–316.

[20] M.F. Desrosiers, Current status of the EPR method to detect irradiated food, Appl. Radiat. Isot. 47 (11) (1996) 1621–1628.

[21] EN 1787, Foodstuffs—Detection of Irradiated Food Containing Cellulose by ESR Spectroscopy, European Committee of Standardization (CEN), Brussels, 2000.

[22] EN 13708, Foodstuffs—Detection of Irradiated Food Containing Crystalline Sugar by ESR Spectroscopy, European Committee of Standardization (CEN), Brussels, 2002.

[23] J.H. Kwon, H.W. Chung, M.W. Byun, ESR spectroscopy for detecting gamma-irradiated dried vegetables and estimating absorbed doses, Radiat. Phys. Chem. 57 (3) (2000) 319–324.

[24] D. Jo, J.H. Kwon, Detection of radiation-induced markers from parts of irradiated kiwifruits, Food. Control. 17 (8) (2006) 617–621.

[25] K. Akram, J.J. Ahn, J.H. Kwon, Characterization and identification of gamma-irradiated sauces by electron spin resonance spectroscopy using different sample pretreatments, Food. Chem. 138 (2) (2013) 1878–1883.

[26] W. Stachowicz, G. Burlińska, J. Michalik, A. DziedzicGocławska, K. Ostrowski, EPR spectroscopy for the detection of foods treated with ionising radiation, in: C.H. McMurray, E. Stewart, R. Gray, J. Pearce, (Eds.), Detection Methods for Irradiated Foods, Current Status, The Royal Society of Chemistry (1996), pp. 23–32. Information Service, Special Publication No. 171.

[27] L. Calucci, C. Pinzino, M. Zandomeneghi, A. Capocchi, S. Ghiringhelli, F. Saviozzi, et al., Effects of γ-irradiation on the free radical and antioxidant contents in nine aromatic herbs and spices, J. Agric. Food. Chem. 51 (4) (2003) 927–934.

[28] E.F.O. De Jesus, A.M. Rossi, R.T. Lopes, Influence of sample treatment on ESR signal of irradiated citrus, Appl. Radiat. Isot. 47 (11) (1996) 1647–1653.

[29] J.J. Ahn, K. Akram, H.K. Kim, J.H. Kwon, Electron spin resonance spectroscopy for the identification of irradiated foods with complex ESR signals, Food Anal. Methods 6 (1) (2013) 301–308.

[30] J.J. Raffi, J.P.L. Agnel, Electron spin resonance identification of irradiated fruits, Int. J. Radiat. Appl. Instrum. Part C. Radiat. Phys. Chem. 34 (6) (1989) 891–894.

[31] Y. Jo, H.K. Kyung, H.J. Park, J.H. Kwon, Irradiated fruits can be identified by detecting radiation-induced markers with luminescence and ESR analyses for different trading fruits, Appl. Biol. Chem. 16 (1) (2016) 1–7.

[32] B.J. Tabner, V.A. Tabner, Electron spin resonance spectra of γ-irradiated citrus fruit skins, skin components and stalks, Int. J. Food Sci. Technol. 29 (2) (1994) 143–152.

[33] E.F. De Jesus, A.M. Rossi, R.T. Lopes, An ESR study on identification of gamma-irradiated kiwi, papaya and tomato using fruit pulp, Int. J. Food Sci. Technol. 34 (2) (1999) 173–178.

[34] Y. Jo, B. Sanyal, H.J. Park, J.H. Kwon, Ethanol extraction-based drying enhanced ESR radical detection in oranges irradiated to different ionizing radiations during storage, Postharvest. Biol. Technol. 112 (2016) 170–175.

[35] H. Delincée, C. Soika, Improvement of the ESR detection of irradiated food containing cellulose employing a simple extraction method, Radiat. Phys. Chem. 63 (3) (2002) 437–441.

[36] M. Korkmaz, M. Polat, Use of electron spin resonance measurements on irradiated sperma lentil seeds to indicate accidental irradiation, Int. J. Food Sci. Technol. 38 (1) (2003) 1–9.

[37] Y. Shimoyama, M. Ukai, H. Nakamura, Advanced protocol for the detection of irradiated food by electron spin resonance spectroscopy, Radiat. Phys. Chem. 76 (11) (2007) 1837–1839.

[38] E.M. Stewart, Detection methods for irradiated foods, Food Irradiat. Principl. Appl. (2001) 347–386.

[39] J. Lee, T. Kausar, B.K. Kim, J.H. Kwon, Detection of γ-irradiated sesame seeds before and after roasting by analyzing photostimulated luminescence, thermoluminescence, and electron spin resonance, J. Agric. Food. Chem. 56 (16) (2008) 7184–7188.

[40] M.F. Desrosiers, Gamma-irradiated seafoods: identification and dosimetry by electron paramagnetic resonance spectroscopy, J. Agric. Food. Chem. 37 (1) (1989) 96–100.

[41] J.J. Raffi, J.P.L. Agnel, L.A. Buscarlet, C.C. Martin, Electron spin resonance identification of irradiated strawberries, J. Chem. Soc. Farad. Trans. 1: Phys. Chem. Condensed Phases 84 (10) (1988) 3359–3362.

[42] N.D. Yordanov, K. Aleksieva, A. Dimitrova, L. Georgieva, E. Tzvetkova, Multifrequency EPR study on freeze-dried fruits before and after X-ray irradiation, Radiat. Phys. Chem. 75 (9) (2006) 1069–1074.

[43] M. Kikuchi, Y. Shimoyama, M. Ukai, Y. Kobayashi, ESR detection procedure of irradiated papaya containing high water content, Radiat. Phys. Chem. 80 (5) (2011) 664–667.

[44] M. Kikuchi, M.S. Hussain, N. Morishita, M. Ukai, Y. Kobayashi, Y. Shimoyama, ESR study of free radicals in mango, Spectrochim. Acta Part A Mol. Biomol. Spectrosc. 75 (1) (2010) 310–313.

[45] N.D. Yordanov, K. Aleksieva, Preparation and applicability of fresh fruit samples for the identification of radiation treatment by EPR, Radiat. Phys. Chem. 78 (3) (2009) 213–216.

[46] N.D. Yordanov, Z. Pachova, Gamma-irradiated dry fruits: an example of a wide variety of long-time dependent EPR spectra, Spectrochim. Acta Mol. Biomol. Spectrosc. 63 (4) (2006) 891–895.

[47] J.J. Ahn, B. Sanyal, K. Akram, J.H. Kwon, Alcoholic extraction enables EPR analysis to characterize radiation-induced cellulosic signals in spices, J. Agric. Food. Chem. 62 (46) (2014) 11089–11098.

[48] J.H. Kwon, H.M. Shahbaz, J.J. Ahn, Advanced electron paramagnetic resonance spectroscopy for the identification of irradiated food, Am. Lab. 46 (1) (2014) 26-+.

[49] K. Malec-Czechowska, G. Strzelczak, A.M. Dancewicz, W. Stachowicz, H. Delincée, Detection of irradiation treatment in dried mushrooms by photostimulated luminescence, EPR spectroscopy and thermoluminescence measurements, Eur. Food Res. Technol. 216 (2) (2003) 157–165.

[50] J. Raffi, M.H. Stevenson, M. Kent, J.M. Thiery, J.J. Belliardo, European intercomparison on electron spin resonance identification of irradiated foodstuffs, Int. J. Food Sci. Technol. 27 (2) (1992) 111–124.

[51] N. Helle, B. Linke, K.W. Bögl, G.A. Schreiber, Elektronen-Spin-Resonanz-Spektroskopie an Gewürzproben, Zeitschrift für Lebensmittel-Untersuchung und Forschung 195 (2) (1992) 129–132.

[52] H. De Cooman, E. Pauwels, H. Vrielinck, E. Sagstuen, F. Callens, M. Waroquier, Identification and conformational study of stable radiation-induced defects in sucrose single crystals using density functional theory calculations of electron magnetic resonance parameters, J. Phys. Chem. 112 (2008) 7298–7307.

[53] Z.M. Da Costaa, W.M. Pontuschkaa, L.L. Campos, A comparative study based on dosimetric properties of different sugars, Appl. Radiat. Isot. 62 (2005) 331–336.

[54] G. Vanhaelewyn, P. Lahorte, F. De Proft, W. Mondelaers, P. Geerling, F. Callens, Electron magnetic resonance study of stable radicals in irradiated D-fructose single crystals, J. Phys. Chem. Chem. Phys. 3 (2001) 1729–1735.

[55] G. Bayram, H. Delincée, Identification of irradiated Turkish foodstuffs combining various physical detection methods, Food. Control. 15 (2) (2004) 81–91.

[56] B. Sanyal, S.P. Chawla, A. Sharma, An improved method to identify irradiated rice by EPR spectroscopy and thermoluminescence measurements, Food. Chem. 116 (2) (2009) 526–534.

[57] G.P. Guzik, W. Stachowicz, J. Michalik, Identification of irradiated dried fruits using EPR spectroscopy, Nukleonika 60 (3) (2015) 627–631.

[58] M.A. Ghelawi, J.S. Moore, R.H. Bisby, N.J.F. Dodd, Estimation of absorbed dose in irradiated dates (*Phoenix dactylifera* L.). Test of ESR response function by a weighted linear least-squares regression analysis, Radiat. Phys. Chem. 60 (1) (2001) 143–147.

[59] R.M. Slave, D. Negut, V.V. Grecu, ESR on some gamma-irradiated aromatic herbs, Rom. J. Phys. 59 (7–8) (2014) 826–833.

[60] M. Braşoveanu, G. Crăciun, E. Mănăilă, E. Ighigeanu, M.R. Nemţanu, M.N. Grecu, Evolution of the levels of free radicals generated on wheat flour and wheat bran by electron beam, Cereal Chem. 90 (5) (2013) 469–473.

[61] A.A. Abdel-Fattah, E.S.A. Hegazy, H.E. El-Din, Radiation-chemical formation of HCl in poly (vinyl butyral) films containing chloral hydrate for use in radiation dosimetry, Int. J. Polym. Mater. 51 (9) (2002) 851–874.

[62] B. Linke, J. Ammon, U. Ballin, R. Brockmann, J. Brunner, H. Delincee, et al., Elektronenspinresonanzspektroskopische Untersuchungen zur Identifizierung bestrahlter getrockneter und frischer Fruchte: Durchfuhrung eines Ringversuchs an getrockneten Feigen und Mangos sowiean frischen Erdbeeren. Report of the Federal Institute for Health Protection of Consumers and Veterinary Medicine. BgW-Heft 03/1996 (Bundesinstitut fur gesundheitlichen Verbraucherschutz und Veterinarmedizin, Berlin), 1996.

[63] J. Raffi, H. Delincee, E. Marchioni, O. Hasselmann, A.-M. Sjoberg, M. Leonardi, et al., Concerted action of the community bureau of reference on methods of identification on irradiated foods. BCR-information: 1994, Commission of the European Communities, (Report EUR/15261/en), Luxembourg, 1994.

[64] B. Linke, N. Helle, J. Ammon, U. Ballin, R. Brockmann, J. Brunner, et al., Elektronenspinresonanz- spektroskopische Untersuchungen zur Identifizierung bestrahlter Krustentiere und Gewurze: Durchfuhrungeines Ringversuches an Nordseekrabben, Kaisergranat und Paprikapulver. Report of the Federal Institute for Health Protection of Consumers and Veterinary Medicine. BgW-Heft 09/1995 (Bundesinstitut fur gesundheitlichen Verbraucherschutz und Veterinarmedizin, Berlin), 1995.

[65] G.A. Schreiber, N. Helle, G. Schulzki, B. Linke, A. Spiegelberg, M. Mager, et al., Interlaboratory Tests to Identify Irradiation Treatment of Various Foods via Gas Chromatographic Detection Of Hydrocarbons, ES, 1996.

[66] C.H. Sommers, X. Fan, Food Irradiation Research and Technology, John Wiley & Sons (2008).

[67] N.D. Yordanov, K. Aleksieva, I. Mansour, Improvement of the EPR detection of irradiated dry plants using microwave saturation and thermal treatment, Radiat. Phys. Chem. 73 (1) (2005) 55–60.

[68] N.D. Yordanov, V. Gancheva, A new approach for extension of the identification period of irradiated cellulose-containing foodstuffs by EPR spectroscopy, Appl. Radiat. Isot. 52 (2) (2000) 195–198.

CHAPTER 6

Electron Paramagnetic Resonance Spectroscopy to Study Liquid Food and Beverages

A.I. Smirnov

North Carolina State University, Raleigh, NC, United States

Contents

6.1 INTRODUCTION

Since ancient times the preservation of food and beverages has been one of the most essential tasks for human society. At the dawn of human civilization, the ability to preserve agricultural products was often a simple issue of survival. Over the course of history, food preservation has always contributed to social and sometimes, even political issues. For example, according to some historians, it was the capacity of ancient Romans to harvest and store grain, and produce olive oil and wine, that gave

Electron Spin Resonance in Food Science.

the society a competitive advantage and propelled the formation of the Roman Empire [1].

The main problem with almost any food is that it would eventually go bad, despite modern technologies and centuries of effort, from ordinary families up to multinational food corporations. Inevitable food degradation and spoilage is a complex process that could involve biological, chemical, and also physical mechanisms. While biological degradation by bacteria and more complex organisms could be solved to a large degree, e.g., by pasteurization and proper packaging, chemical degradation is the most difficult to address because of the complexity of possible chemical reactions, with many of them being oxidative.

Oxidation is the most common type of chemical reaction that would lead to the degradation of food and beverage products [2]. Typically, the exposure of a food or beverage to oxygen would trigger a chain of chemical reactions involving proteins, pigments, fatty acids, and lipids producing other compounds with undesirable biochemical properties (i.e., toxicity), taste, smell, and color. Some of these reactions could be induced by exposure to light (i.e., photosensitized oxidation), heat (i.e., thermal oxidation), or mediated by enzymes (i.e., enzymatic oxidation). Many of these processes occur via free radical mechanisms and involve chain reactions.

Electron paramagnetic resonance (EPR) spectroscopy is a versatile and exceptionally sensitive technique for analyzing molecular systems possessing unpaired electronic spins, such as paramagnetic metal ions and organic free radicals. While EPR can be directly applied to detect unpaired electron species in solid food products (these applications have been extensively reviewed in other chapters of this book), liquids and beverages present an additional challenge to the method because of the generally short lifetime of free radicals that result in an insufficient concentration of radicals for direct EPR detection. Indeed, reactive oxygen species (ROS) and intermediate organic free radical compounds that could be intimately involved in food oxidation processes are generally highly reactive and exist only transiently—just until the next molecular collision with other organic molecules. Nevertheless, some of the organic radicals and, particularly, phenolic compounds, could be accumulated in liquid food products in concentrations sufficient for EPR detection. Paramagnetic metal ions and their complexes can also be readily detected by EPR in some beverages. However, the former examples generally represent exceptions to the rule, as the vast majority of liquid food products and beverages remain "EPR-silent." The absence of an EPR signal does not necessarily mean the lack of processes involving ROS and/or other free radicals, but rather the short lifetime of the radical species formed. The problem of the short lifetime, and the resulting low steady-state concentrations, can be overcome by trapping the radicals with specially designed compounds known as spin-traps. The spin adducts formed in the course of spin-trapping reactions can then be characterized by EPR (e.g., Refs. [3,4]). The EPR spin-trapping method could also be employed to characterize the radical scavenging activity of other compounds (i.e., antioxidants) by artificially generating ROS and observing the intensity of the trapped EPR signals (e.g., Ref. [5]). Alternatively, the antioxidant properties of beverages could be probed by observing the scavenging of free radicals and other redox-active compounds by nitroxides (e.g., Ref. [6]) and other stable free radicals such as DPPH (1,1-diphenyl-2-picryl-hydrazyl) [7,8].

Finally, but not last, EPR is uniquely positioned for studying the mechanisms of chemical reactions involving free radicals. For such studies, short-lived free radical intermediates could be stabilized by rapid freezing (the freeze-quench method) immediately after initiation. Another method involves time-resolved electron paramagnetic resonance (TREPR) spectroscopy of short-lived radicals generated after a laser pulse [9]. The latter method was particularly useful for unraveling a molecular mechanism leading to the formation of an undesirable "lightstruck" flavor in beer exposed to light [10].

The main objective of this chapter is to discuss the applications of EPR to the study of beverages, as well as other liquid food products. The focus will be on emphasizing different approaches from the observation of endogenous unpaired electronic spin species in beverages to study their redox and antioxidant properties and free reduction mechanisms. We shall also highlight EPR studies of beer as an example of a now widespread industrial application [11–13] of this powerful and sensitive magnetic resonance spectroscopy.

6.2 ENDOGENOUS UNPAIRED ELECTRONIC SPIN SPECIES IN BEVERAGES

6.2.1 Soluble Organic Free Radicals

Many of the chemical reactions involving free radicals are diffusion-controlled and therefore, the steady-state concentration of ROS in beverages at ambient conditions must remain low, certainly below the nanomolar levels accessible by conventional research-grade continuous wave (CW) EPR spectrometers. For these reasons, essentially all but a very few beverages remain EPR silent, with no signals detected even though free radical reactions may still be occurring continuously. One example of the latter is bottled beer [14]. Nevertheless, in some beverages, persistent organic free radicals still could be detected and characterized by EPR.

One example of organic free radicals present in beer was provided by Jehle et al., who described the room temperature EPR detection of an exceptionally stable organic radical(s) directly in dark beers and in sweet wort produced with dark malts [15]. The detected X-band (9.5 GHz) EPR spectrum consisted of a broad (ca. 28 G peak-to-peak) single line at $g \approx 2.0040$, with no even partially resolved hyperfine components (Fig. 6.1). Such spectra are typical for other carbon-centered radicals when recorded at X-band EPR frequency. The EPR signals were detected at the initial step of the mashing process, and its intensity increased when the fraction of dark malt was increased. While no EPR spectra of the roasted malt alone were provided by the authors, the data suggested that the radical originated from a water-soluble melanoidin-derived complex that is formed during the malt roasting [15].

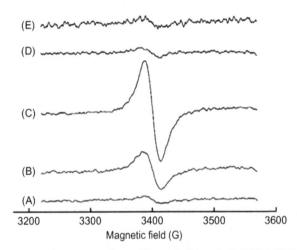

Figure 6.1 Room temperature air-equilibrated X-band (9.5 GHz) EPR spectra of: (A) stout beer; (B) wort containing 25% dark malt; (C) wort containing 100% dark malt; (D) the first colored fraction from a gel filtration; and (E) wort containing 100% dark malt recorded under an N_2 atmosphere. *Reproduced with a permission from D. Jehle, M.N. Lund, L.H. Ogendal, M.L. Andersen, Characterisation of a stable radical from dark roasted malt in wort and beer, Food. Chem. 125 (2) (2011) 380–387.*

Jehle et al. also performed additional characterization of the radical complex using different precipitation protocols, size-exclusion chromatography, and light scattering experiments that led them to suggest a melanoidin-derived radical associated with a high molecular weight complex of approximately 10^6–10^8 g/mol [15]. The radical proved to be exceptionally stable, as the EPR signal was not affected by heating to ca. 70°C. In addition, hydroxyl radicals produced in the course of the Fenton reaction initiated by the addition of up to 0.5 mM H_2O_2 alone or in combination with $FeSO_4$, did not affect the observed signal. While such exceptional stability of the free radical in an oxidizing environment could serve as an indication of complete steric protection and decoupling of the radical from redox processes, the EPR signal from a wort made from 100% dark malt was observed to decrease to about half of the initial intensity upon equilibration of the wort with nitrogen without any changes in the line shape (Fig. 6.1E). The change was fully reversible upon re-exposure of the sample to air [15]. This observation could serve as an indication of at least some involvement of the observed radical in the oxidative reactions, even though its direct relation to the enhanced oxidative stability of dark versus light beers still remains unclear.

Brewed coffee and coffee beans represent, perhaps, the best known example of food exhibiting exceptionally stable organic radicals that give rise to EPR spectra (e.g., see Refs. [16–20]). Green coffee beans before roasting yield a somewhat weak single-line X-band EPR spectra at g_{iso}=2.0045 ± 0.0002 with a peak-to-peak linewidth of $\Delta B_{p\text{-}p}$=6.7 ± 0.6 G [17]. During the roasting, the coffee beans accumulate much more intense free radical signals that change the width and g-factors during the roasting process, based on an EPR study that closely followed industrial roasting conditions [19]. Another recent EPR examination of roasted coffee beans sourced from an industrial coffee roasting plant identified at least three different organic radical species [20]. Continuous wave EPR spectra collected from *Coffea arabica* (Arabica) beans sourced from Brazil and roasted for 2, 4, 6, 8, 10, and 12 min at temperatures up to 220°C exhibited changes in g-factors and linewidth that were related to the formation and disappearance of different radicals during the roasting process [20]. Similar observations were also made previously [19]. Notably, the EPR parameters $\Delta B_{p\text{-}p} \approx 6.8$ G and $g_{iso} \approx 2.004$ of the remaining free radicals, after completing roasting and storing the coffee beans, were essentially the same among the studies conducted by different groups and at different times [17,19,20], and were similar to those of unroasted coffee beans, although signal intensity increased by more than 30-fold [17]. These observations indicate the formation of radicals with very similar EPR spectra, regardless of the origin of the coffee and variations in roasting conditions. Troup et al. also observed an effect from the grinding of coffee beans on the intensity of the observed EPR signals, and suggested that organic free radicals may not be distributed uniformly across the coffee beans [20]. Indeed, EPR imaging experiments of the individual beans reported much earlier revealed the heterogeneity of the free radical distribution [17].

Free radicals formed in the roasted coffee beans are exceptionally stable and are also found to be present in brewed coffee, together with characteristic EPR spectra from Fe^{3+} and Mn^{2+} ions (e.g., Refs. [18,20]). Fig. 6.2

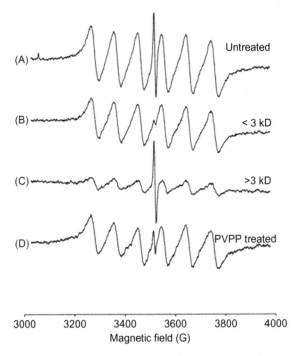

Figure 6.2 Room temperature X-band EPR spectra of brewed dark roast coffee after various treatments: (A) filtered through a low-binding 0.45 µm pore size syringe filter to remove particulates; (B) filtrate from the first pass through an Amicon 3 kDa cutoff centrifugal filtration cartridge; (C) retentate after three passes through the same cartridge; and (D) unfiltered brew treated with polyvinylpolypyrrolidone (PVPP; Fluka), which adsorbs phenolic compounds. *Reproduced with a permission from G.J. Troup, L. Navarini, F.S. Liverani, S.C. Drew, Stable radical content and anti-radical activity of roasted Arabica coffee: from in-tact bean to coffee brew, PLoS One 10 (4) (2015) e0122834.*

shows an example of a room temperature X-band EPR spectrum from a coffee brewed from dark roasted beans that exhibits a sharp central line attributed to a water soluble organic free radical together with six hyperfine components arising from Mn^{2+} aqua ions [20]. The concentration sensitivity of modern research-grade EPR spectrometers, such as the E500 X-band instrument outfitted with a Bruker super-high-Q probe head ER 4122SHQE (all from Bruker Biospin, Karlsruhe, Germany), was sufficient to determine the concentration of Mn^{2+} in the brew to be $\approx 13\,\mu M$, and that of the stable organic free radicals up to $\approx 15\,nM$ [20]. The width of the single line free radical components was $\approx 5\,G$, with $g_{iso} = 2.0039$—magnetic parameters that are consistent with the predominant free radical species formed in coffee beans upon roasting. To further confirm that the organic free radicals extracted by the hot water brewing process (5 min at $92 \pm 1°C$) were indeed associated with soluble compounds rather than the remaining coffee particulates formed upon fine grinding of the beans, Troup et al. employed a low-binding $0.45\,\mu m$ pore size syringe filter (Millipore). By carrying out consequent ultrafiltration with Amicon (Millipore) 3 kDa cutoff centrifugal filtration cartridges, and also treating the unfiltered brew with polyvinylpolypyrrolidone that adsorbs phenolic compounds, it was determined that the organic radical EPR signal was primarily associated with high molecular weight (HMW) phenolic compounds present in the coffee brew [20]. The same HMW fraction also provided the higher color intensity at 420 nm that is indicative of melanoidins—compounds generated in the course of the Maillard reaction from reducing sugars and amino acids. These observations let Troup et al. to speculate that the EPR-active radical species in brewed coffee reside within the aromatic groups within the melanoidins [20]. They noted that caramelization of sugars occurring in the course of production of a number of food flavorings and additives would also give rise to free radical EPR signals with similar line width and g-factors, but those radicals are not associated with the phenolic compounds exhibiting antioxidant properties [21].

EPR signals associated with phenolic compounds were also detected in red wine [22]. The signals have virtually the same $g_{iso} = 2.0038 \pm 0.0001$ as other carbon-based radicals described in this section, but a narrower linewidth of only $\Delta B_{p-p} = 2.0 \pm 0.1\,G$. The phenolic origin was confirmed by treating wine with polyvinylpolypyrrolidone. Furthermore, the radicals were detected in red wine, whether fermented in oak or not, and in white wine but only after oak fermentation. The latter indicates that a fraction of the phenolic compounds in wine could be originating from oak casks employed first in fermentation and then in wine aging [22]. Polyphenolic compounds present in either red wine or red wine extracts (primarily, these compounds are found in grape skins and seeds [23]) are considered to be potent antioxidants, and have been shown to be effective in reducing myocardial ischemic reperfusion injury [24].

Other beverages prepared from plants with a high content of phenolic compounds are expected to give rise to EPR spectra similar to those detected in red wine. For example, detailed X-band EPR studies of different teas revealed the presence of both Mn^{2+} and a single line free radical signal of polyphenolic origin (e.g., see Refs. [25–28]).

6.2.2 Paramagnetic Metal Ions

In addition to free radicals, EPR spectroscopy is widely used to study paramagnetic metal ions and their complexes. EPR is particularly sensitive to Cu^{2+}, Mn^{2+}, and Fe^{3+} ions because of the favorable electronic relaxation times. The latter two represent the 12th and the 5th most abundant elements in the earth's crust [29], and are commonly present in plants and plant extracts. Mn^{2+} is particularly easy to detect and identify by EPR even at room temperature, because of a characteristic six-line hyperfine pattern (cf. Fig. 6.2) due to an interaction of the electronic spin with ^{55}Mn nuclear spin $I = 5/2$.

The EPR spectra of Mn^{2+} are sensitive to the ligand field that changes the magnetic parameters (i.e., zero-field splitting or ZFS) of this electronic $S = 5/2$ spin that manifests in different linewidths and the appearance of additional lines between the main six hyperfine components. The additional weaker lines are due to the so-called forbidden hyperfine transitions that are partially allowed at low magnetic fields corresponding to X-band EPR. The ZFS effects on Mn^{2+} spectra could be very useful for effectively "removing" the EPR signal of unchelated Mn^{2+} from an aqueous sample by adding the chelating agent DTPA (diethylenetriaminepentaacetic acid), which binds to many di- and trivalent metal ions with a high affinity, and which has an X-band EPR spectrum that is too broad to be detected at room temperature [20].

The EPR spectra of Mn^{2+} are sensitive to the ligand field and these effects could be modeled by computer simulations to provide additional insights into the structure of Mn^{2+} complexes extracted from tea leaves (e.g., Ref. [25]). One of the problems with naturally occurring Mn^{2+} complexes is their heterogeneity, as some of the Mn^{2+} ions could be bound to catechins or polyphenolic compounds, giving rise to some broad EPR spectra at X-band [28]. The fitting of such multicomponent EPR spectra would be more difficult than analyzing specific Mn^{2+} complexes and/or protein binding sites in which paramagnetic metal ions could exist in only one or two configurations. For these reasons, similar to the analysis of free radicals, the researchers typically combine EPR and separation technologies such as an ultrafiltration and/or binding to water-insoluble polyvinylpolypyrrolidone, and analyze the obtained fractions for EPR signals (e.g., see Ref. [20]).

Another paramagnetic metal ion commonly present in plant extracts is Fe^{3+} that gives rise to an EPR line at $g \approx 4.3$ (e.g., Ref. [28]). Such spectra could be readily analyzed using available data on the effects of the ligand field on EPR parameters (e.g., Refs. [30,31]). The presence of unchelated Fe^{3+} as well as Cu^{2+} in food and beverages could be undesirable, because these transition metal ions could play the role of catalysts in the Fenton reaction, producing free radicals (e.g., Ref. [32]) that would eventually lead to food spoilage. One example was provided by Kaneda et al. who employed CW X-band EPR at $77\,K$ to study the role of metal ions in beer deterioration [33]. Specifically, these authors have followed a $g = 4.3$ EPR signal due to non-heme iron in the course of beer oxidation that was accelerated by the addition of H_2O_2, Fe^{3+}, or Fe^{2+} (compounds responsible for Fenton's reaction). During the oxidation the amplitude of the $g = 4.3$ Fe^{3+} EPR signal increased, while the oxidative degradation of the main bitter components, isohumulones, was accelerated [33]. It was suggested that during storage Fe^{2+} ions dominant in fresh beer are oxidized to $Fe^{3+}$and, consequently, form complexes with other beer components (non-heme Fe^{3+}) leading to haze formation [33]. An observation of an inhibiting effect of iron-chelating compounds DETAF'AC or EDTA on oxidative beer degradation was in full agreement with this hypothesis [33].

Changes in iron oxidation state have also been observed in brewed tea. For example, the Fe^{3+} EPR line at $g = 4.32$ in black tea has been effectively quenched by adding glucose, fructose, maltose, and lactose, likely because of reducing Fe^{3+} to diamagnetic Fe^{2+} by transferring or sharing an electron from a hydroxyl group in the sweeteners [28]. At the same time, the white sugar (sucrose) quenches the Mn^{2+} and the free radical EPR signal in black tea, but not the Fe^{3+} EPR line. The significance of these finding for nutrition is uncertain, because the exact relationship between the free radical signal, metal complexes, and the antioxidant activity of black tea is not fully clear at this moment [28].

Finally, as was noted by Biyik and Tapramaz, the EPR spectra of black tea are not fully characteristic for this beverage, as some riverine materials from tropical ecosystems and municipal solid waste composts show almost the same EPR components [28]. This is not surprising, because the soil samples are known to contain both Mn^{2+} ions and also humic substances—the major components of natural organic matter. The latter could contain stable free radicals with concentrations up to 10^{18} spin/g, exhibiting a single line EPR spectrum of about 6 G wide at X-band [34].

6.3 REDOX AND ANTIOXIDANT PROPERTIES OF BEVERAGES BY ELECTRON PARAMAGNETIC RESONANCE

6.3.1 Spin-Trapping Electron Paramagnetic Resonance Studies of Beverages

It is well established that the chemical mechanisms of food spoilage could involve reactive oxygen species (ROS). The ROS are exceptionally reactive chemically and, therefore, are only present in miniscule concentrations that are not sufficient for direct detection by conventional EPR methods. This problem can be overcome by stabilizing the free radical intermediates by reacting ROS with specially designed compounds known as spin–traps [3]. Fig. 6.3 shows examples of diamagnetic compounds that are commonly used for trapping reactive radicals to form more stable

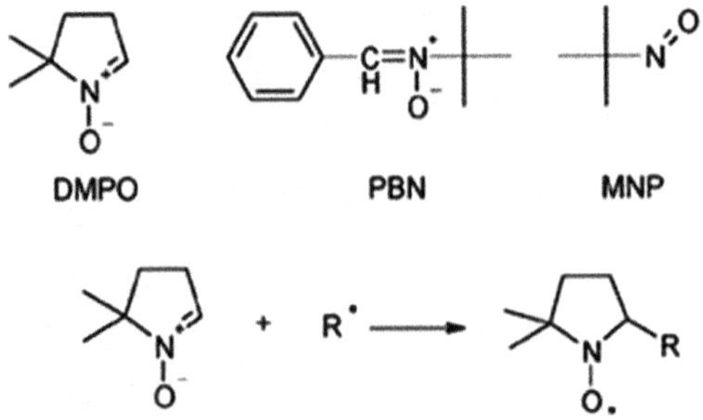

DMPO   PBN   MNP

Figure 6.3 Top: the chemical structure of DMPO (5,5-dimethyl-1-pyrroline-*N*-oxide), PBN (*N-tert*-butyl-α-phenylnitrone), and MNP (2-methyl-2-nitrosopropane) spin-traps. Bottom: a sample of chemical trapping of a reactive free radical resulting in the formation of a spin adduct that is more stable due to steric protection. *Reproduced with permission from M. Conte, H. Miyamura, S. Kobayashi, V. Chechik, Enhanced acyl radical formation in the Au nanoparticle-catalysed aldehydeoxidation. Chem. Commun. 46 (1) (2010) 145–147 [39].*

paramagnetic compounds (i.e., spin adducts). The latter could consequently be detected and characterized by EPR (e.g., Refs. [3,35]), NMR [36], or mass spectrometry [37,38].

The identification of spin adducts by EPR is typically carried out by analysis of spectra recorded in liquids. Fig. 6.4 shows an example of an experimental room temperature X-band EPR spectrum of PBN spin adducts in beer. Such spectra are characteristically split into multiple components by isotropic hyperfine interactions of the electronic spin $S = 1/2$ with the neighboring magnetic nuclei, while any anisotropic hyperfine interactions are averaged out by a rapid tumbling of these small molecules in non-viscous fluids.

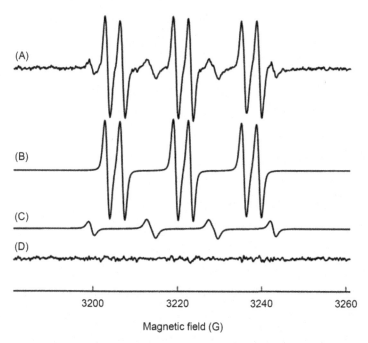

Magnetic field (G)

Figure 6.4 (a) A typical experimental room-temperature X-band EPR spectrum of PBN spin-adducts formed in a lager beer after accelerated oxidative stress at 60°C. (b) Results of the least squares simulations of the main spin adduct signal using software described earlier [50] yielded magnetic parameters $a_N \approx 15.8\,G$ and $a_H \approx 3.5\,G$ that are consistent with either hydroxyl-PBN or 1-hydroxyethyl-PBN spin adducts. (c): Simulated spectrum of a minor PBN spin adduct with magnetic parameters $a_N \approx a_H \approx 14.4\,G$. (d). The fit residual—a difference between the experimental and simulated spectra shows no systematic deviations between the model and the experiment. See text for further details.

The isotropic hyperfine coupling constants are characteristic and tabulated for many spin-trap–free radical combinations (e.g., Ref. [35]). In some cases additional information could be obtained from g-factor shifts and/or the effects of molecular motion on the EPR line shapes measured at high magnetic fields/high frequencies, as was demonstrated by Smirnova et al. [40] on examples of PBN spin adducts measured at W-band (94 GHz, which corresponds to a c.3.3T resonant magnetic field for $g \approx 2$ EPR species).

Perhaps the first demonstration of the use of spin-trapping EPR methods to characterize common beverages such as beer was provided by Kaneda et al. [41]. The authors incubated bottled non-pasteurized lager beer with a spin-trap PBN (see Fig. 6.3 for PBN structure) at 60°C, and observed the formation of spin adducts that were consistent with being formed by hydroxyl radicals [41]. Since then, spin-trapping EPR tests have been widely employed to study the formation of ROS in beer subjected to an oxidative stress—typically, heating to ca. 50–60°C in the presence of atmospheric oxygen (e.g., Refs. [13,42–48]).

Fig. 6.4 shows a typical experimental room temperature X-band EPR spectrum of PBN spin adducts formed in a lager beer after accelerated oxidative stress at 60°C. The signal is dominated by six lines of almost equal peak-to-peak intensity that could be modeled by electronic spin $S = 1/2$ species exhibiting isotropic hyperfine interactions with one nuclear spin $I = 1$ (corresponding to ^{14}N, $a_N = 15.8$ G), and one $I = 1/2$ spin (^1H, $a_H = 3.5$ G) [48]. Least-squares fitting of the entire spectrum shown in Fig. 6.4A also reveals the presence of a minor spin adduct with $a_N \approx a_H \approx 14.4$ G (Fig. 6.4C). The latter signal likely originates from a hydrolysis of PBN adducts [49], and is not always present. Typically, the researchers are focusing on the analysis of the main six-line component (e.g., Ref. [13,42–48]).

The EPR hyperfine a_N and a_H parameters of the dominating main six-line component of the EPR spectrum in Fig. 6.4A are consistent with those of a PBN hydroxyl spin adduct (e.g., Ref. [51]), and this was the original interpretation by Kaneda et al. [41]. It should be noted that even if the initial concentration of PBN in spin-trapping experiments is relatively high, typically up to 50 mM, it is still much lower than the about 1 M concentration of ethanol present. While the reaction rate for trapping hydroxyl radicals by PBN is relatively high (from 6×10^9 to 9×10^9 M/s [52]), much higher concentrations of ethanol in beer would lead primarily to the formation of the 1-hydroxyethyl radical first that would eventually be trapped by PBN [45]. Based on the available data for the rate constants of possible bimolecular reactions and a typical ethanol concentration of 1 M, it was concluded that the ratio of the 1-hydroxyethyl PBN to hydroxyl PBN adducts under these conditions could be 4:1 [48]. Differentiating between such adducts from X-band EPR spectra is not easy because of the overlapping components with very similar isotropic hyperfine parameters: $a_N = 15.94$ G and $a_H = 3.34$ G for 1-hydroxyethyl-PBN [51], and $a_N = 15.7$ G and $a_H = 3.2$ G for hydroxyl-PBN [41]. Additional adducts formed by the secondary radicals generated in the chain propagation steps (e.g., see Ref. [45]) can be neglected because these radicals are less reactive than $HO^{\bullet}$ and are typically present in much lower concentrations.

As was initially reported by Uchida et al. [43,44], at the beginning of PBN spin-trapping experiments the amplitude of EPR signals from spin adducts is low or is not detectable at all (e.g., Ref. [48]). However, after a certain time interval the formation of spin adducts would accelerate (cf. Fig. 6.5) and eventually reach a plateau [48]. In chemical kinetics an initial slow stage of a chemical reaction is called the induction period, but in the brewing industry and food science it is also called a lag period.

Beer staling is thought to be a complex oxidation process that is accelerated by increasing temperature, the presence of oxygen, and several other factors. Even when the oxygen concentration is low, a chain of oxidation

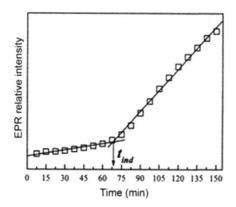

Figure 6.5 Typical time evolution of the relative intensity of EPR spectra during a thermally accelerated aging of an oxygenated beer sample at 60°C in the presence of a PBN spin-trap. The induction period is indicated by an arrow. *Reproduced with permission from V. Brezova, M. Polovka, A. Stasko, The influence of additives on beer stability investigated by EPR spectroscopy. Spectrochim. Acta Part A Mol. Biomol. Spectrosc. 58 (6) (2002) 1279–1291.*

reactions could be initiated by HO$^\bullet$ radicals. The latter are produced in the course of Fenton's reaction which catalyzes H_2O_2 decomposition by transition metal ions, typically iron and/or copper that could be present in beer. Alternatively, HO$^\bullet$ radicals could be formed in the course of thermally or photochemically induced homolysis of some weak molecular bonds of other organic components of beer. As already mentioned, the high chemical reactivity of the hydroxyl radicals—one of the highest among ROS—would result in the formation of additional radical adducts upon HO$^\bullet$ collisions with almost any organic molecules present in a solution. In beer, the most abundant organic compound is ethanol, which at ≈ 1 M concentration exceeds that of carbohydrates, proteins, amino acids, etc., significantly. These considerations provide a strong argument that in lager beers the hydroxyl radical is almost immediately converted into 1-hydroxyethyl radicals with a pseudo-first-order rate constant of $1.9\times 10^9\,s^{-1}$, causing a chain of free radical reactions yielding a number of products [45]. Initially, these radicals will be effectively neutralized by a pool of natural antioxidants present in beer. Once this pool is exhausted, the main reaction pathway switches to the formation of EPR-active spin adducts, and this moment in time corresponds to the above mentioned induction or lag period (cf. Fig. 6.5).

Previously, it was suggested that the lag period could be used as an indicator of beer flavor stability, and it was found to correlate with beer age [11–13,44,53,54]. The EPR lag period compares well with other indexes that were put forward to evaluate the aging of bottled lager beer [55]. Further analysis depends upon kinetic models describing free radical processes in beer. One of these models has been described by Kocherginsky and coworkers, who demonstrate that the total kinetics of the process can be characterized by three parameters: the lag period, the rate of spin-trap adduct formation and, finally, the steady-state spin adduct concentration [48]. These authors also suggested a new dimensionless parameter to characterize the antioxidant pools—a product of the lag time and the rate of spin-trap radical formation immediately after the lag time, normalized by the steady-state concentration of the spin adducts [48].

Spin-trapping EPR studies of free radicals formed in beer under aerobic conditions have been expanded to traps other than PBN in order to provide a better characterization of the intermediate ROS species [45,56,57]. For example, Andersen and Skibsted employed another common spin-trap, DMPO (2-methyl-2-nitrosopropane, 5,5-dimethyl-1-pyrroline N-oxide), to identify the radicals formed during the beer oxidation process and, specifically, to show the primary involvement of the 1-hydroxyethyl radical [45]. The effects of several additives such as hexamethylenetetramine or sulfite during the mashing step of beer preparation [58], riboflavin and riboflavin binding proteins [56], coloring agents (e.g., those employed in specialty malts as well as artificial caramel colorant), [59] and antioxidant vitamins E and C [60] were also investigated.

Spin-trapping EPR methods are also well suited to studies of other liquid food products and beverages, as was demonstrated by several reports on, e.g., tea [61], food lipids [62] and vegetable oils [63], and wine [64–67]. In all the latter studies spin-trapping EPR could be applied to identify endogenous radicals, typically under oxidative stress conditions imposed by exposing the sample to oxygen and/or elevated temperatures. Such conditions are required to accelerate the formation of ROS. Otherwise, the production of free radicals would be too slow for EPR detection. An alternative approach for increasing ROS concentrations would be to generate those exogenously, e.g., by adding components of Fenton's reaction (typically, by mixing Fe^{2+} and H_2O_2) to produce hydroxyl radicals. This and other ROS generating systems have been reviewed elsewhere (e.g., Ref. [68]). Then ROS scavenging activity of beverages and/or specific compounds could be assessed by setting up a competition assay between potential antioxidants and spin-traps with paramagnetic spin adducts being detected by EPR (e.g., Refs. [5,69]), as well as other methods (e.g., Refs. [69,70]). One example of using a number of ROS generating systems and the consequent EPR spin-trapping has been provided by Yen and Chen, who reported on the antioxidant activity of various tea extracts (green tea, pouchong tea, oolong tea, and black tea) toward specific ROS [71].

Overall, spin-trapping EPR has proven to be an exceptionally sensitive and informative method of both identifying ROS in liquid food and beverages, as well as for assessing the antioxidative activity of natural compounds and food additives.

6.3.2 Antioxidant Properties of Beverages Assessed by Spin Probe Electron Paramagnetic Resonance

While spin-trapping EPR experiments are capable of providing a wealth of information about the ROS that are present, it is not the only way to assess effective concentrations of available antioxidants (i.e., the antioxidant pool). A different, and at the same time simpler, approach was first outlined by Blois back in 1958: "since a most interesting role of antioxidants … is their interaction with oxidative free radicals, it appeared of interest to carry out their determination in a like manner" [72]. For this purpose, Blois suggested monitoring the concentration of DPPH (2,2-diphenyl-1-picrylhydrazyl), "which is generally available in laboratories in which electron spin resonance experiments are conducted" [72]. DPPH is, perhaps, the best known example of a stable free radical in both solid form and in solutions. When dissolved in methanol, it appears purple in color by absorption at 515 nm. The solution color changes from purple to yellow when DPPH˙ accepts a hydrogen atom from a scavenger molecule (i.e., antioxidants), forming a diamagnetic compound DPPH2. This reaction is the basis for a DPPH antioxidant assay that consists of monitoring DPPH˙ concentration by either EPR or a spectrophotometer [7,72,73]. While spectrophotometers for monitoring color change are available in many labs, the EPR method of quantifying the DPPH˙ signal is more broadly applicable, as it can be used for turbid and/or opaque solutions that may be difficult to examine with optical methods. Thus, EPR is now widely used for monitoring DPPH˙ concentration and evaluating antioxidants in tea [71] and coffee [74], other extracts of natural products [75], beer [8,76], wine [77] and several other commercially available alcoholic beverages [78], fruit juices, drinks and nectars [79], and other food products, such as, e.g., Slovak honey [80].

The DPPH antioxidant assay has been in use for almost 60 years; however, it is not without some disadvantages. Firstly, DPPH is not soluble in water and, therefore, this assay is typically performed in methanol or 1:1 (v/v) methanol−water solutions. Clearly, water−methanol mixtures do not mimic the aqueous phase very well, and could potentially alter the reaction pathways because of the high methanol concentration. Furthermore, as discussed by Mishra and coworkers, different antioxidants might react with DPPH with different kinetics, or might not react at all [73]. For example, when discussing some contradictory results of DPPH assays for aged beer, and comparing those with another common redox indicator DCPI, Kaneda and coworkers suggested that DPPH reacts mainly with polyphenols, while DCPI is more selective for the products of polyphenol oxidation, and also reductones and melanoidines [81]. Finally, an EPR-silent DPPH2 adduct could also reoxidize back to the EPR-active DPPH˙ form, thus complicating data interpretation [73].

While the main advantage of DPPH as a redox indicator is the wide availability of spectrophotometers for monitoring color change, it is not the only stable free radical available for the experiments initially

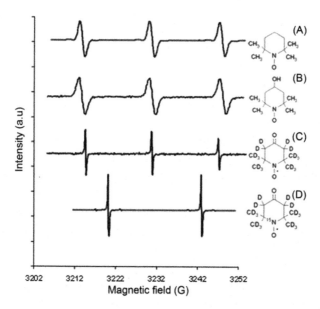

Figure 6.6 The structure and solution EPR spectra of common nitroxides employed by Kocherginsky et al. to assess the "reducing power" of beer: (A) Tempo (2,2,6,6-tetramethylpiperidine1-oxyl); (B) Tempol (4-hydroxy-Tempo); (C) perdeuterated-Tempone (4-oxo-Tempo-d_{16}); (D) ^{15}N-substituted perdeuterated-Tempone (4-oxo-Tempo-d_{16}-^{15}N). *Reproduced with minor modifications and permission from N.M. Kocherginsky, Y.Y. Kostetski, A.I. Smirnov, Use of nitroxide spin probes and electron paramagnetic resonance for assessing reducing power of beer. Role of SH groups. J. Agric. Food. Chem. 53 (4) (2005) 1052–1057.*

described by Blois [72] if one employs EPR detection instead. One wide class of stable organic free radicals that fit this purpose is nitroxides (cf. Fig. 6.6). The organic chemistry of nitroxides is well developed and could be tailored for specific applications in materials science, chemistry, biology, biophysics, and experimental medicine [82]. The nitroxides have many useful and unique applications in analytical chemistry as spin probes when these molecules are introduced in small quantities solely for the purpose of gaining information on the system under study [83]. The EPR spectra of nitroxides are known to be exceptionally sensitive to the physical properties of the microenvironment, such as viscosity, polarity, and the presence of other paramagnetic [84] and redox species (e.g., reviewed in Ref. [85]). The latter applications are based on the chemical reaction of a nitroxide R–NO$^{\bullet}$ with a reducing agent [H] to produce the corresponding EPR-silent hydroxyl amine that can be reoxidized back by an oxidizing agent [O]:

$$R-NO^{\bullet} \underset{[O]}{\overset{[H]}{\rightleftarrows}} R-NOH \tag{6.1}$$

These chemical reactions can be readily monitored, even in non-transparent media, by detecting the EPR signals of R–NO$^{\bullet}$. Whether or not a nitroxide would act as an electron acceptor depends on the relative redox potentials of R–NO$^{\bullet}$ and [H], and the redox potential of nitroxides is known to be affected by their chemical structure (e.g., see Ref. [86]). The standard redox potentials are equally important parameters of other indicators and should be considered for the interpretation of data from antioxidant assays. We note that the standard redox potentials of antioxidant indicators DPPH and DCPI are 1.2 and 0.67 V, respectively, while the redox potential of Tempo is about 0.4 V [87].

Nitroxides have been used to monitor free radical processes in beverages in the past. For example, Brezova and coworkers employed nitroxide Tempol (cf. Fig. 6.6B) to monitor aging processes in beer, and correlated trends in kinetics in Tempol reduction with beer age and the lag period measured with the spin-trap PBN [47]. These authors assumed that the main reactions leading to the loss (or reduction) of Tempol EPR signal are those that terminate free radicals generated in beer.

Kocherginsky et al. carried out detailed kinetic studies of reactions of several nitroxides in fresh and aged beer [6]. An important feature of these studies was the use of gas-permeable Teflon (poly(tetrafluoroethylene), PTFE) capillaries (0.81 mm i.d., 0.86 mm o.d.; Zeus Industrial Products, NJ) that allowed for a fast (i.e., within 3–4 min) re-equilibration between the gas in the capillary surrounding and the one dissolved in a liquid drawn inside such a capillary (see also Ref. [88]). Thus, the redox reactions with participation of nitroxides could be carried out at controlled concentrations of molecular oxygen [6]. Removal of molecular oxygen by equilibrating the capillary with nitrogen would not only mimic the optimal storage conditions (i.e., beer in a hermetically sealed can or bottle), but would also prevent the reverse reaction [1] that would complicate the data analysis.

Another important feature of these experiments was the use of isotopically substituted nitroxides to improve the signal-to-noise ratio of

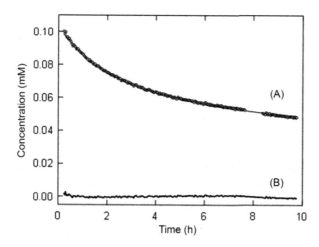

Figure 6.7 (A) Typical kinetics of reduction of 4-oxo-Tempo-d_{16} in nitrogen-equilibrated Miller Genuine Draft beer measured by X-band EPR at room temperature (open circles) superimposed with least-squares to a kinetic model described in the text. The initial concentration of nitroxide was 100 μM and at the end of the experiment the nitroxide concentration was reaching a plateau. (B) The fit residual—a difference between the experimental data points and least-squares fit.

the EPR spectra owing to a substantial line narrowing (cf. Fig. 6.7). For example, the peak-to-peak linewidth of each of the nitrogen hyperfine components of aqueous 4-oxo–Tempo spectrum at X-band is about 1.1 G, primarily due to unresolved hyperfine coupling to several ^{1}H nuclear spins. Deuteration decreases the magnitude of this coupling and causes the line to narrow down to ca. 0.25 G, with further narrowing obtained by deoxygenation. The resulting gain in the signal peak-to-peak amplitude allows the nitroxide concentration to be decreased to 0.1–1.0 μM or even lower. At such concentrations any reducing species [H] present in beer are expected to be in large excess to the nitroxide, and its addition is not expected to change the pathways of redox reactions. Indeed, when 4-oxo-Tempo-d_{16} was added at an initial concentration of even 10 μM to deoxygenated Miller Genuine Draft (MGD) beer, the reaction proceeded at an effective first order level to complete exhaustion of the nitroxide [6]. Once all the nitroxide was depleted, re-equilibration of the solution with oxygen resulted in some recovery of the EPR signal, demonstrating that at least one of the reaction products was the corresponding hydroxylamine [6,87].

Kocherginsky et al. also varied the nitroxide concentration to determine the reaction mechanism, and to deduce the effective concentration of the reducing agents in beer and the corresponding reaction rate constant. Fig. 6.7 shows the results of a typical nitroxide reduction experiment carried out with a higher 100 μM concentration of 4-oxo-Tempo-d_{16} added to nitrogen-equilibrated MGD beer at room temperature. The nitroxide concentration was determined by least-squares fitting of the spectra that allows for accurate determination of the EPR double-integrated intensity and accounts for linewidth effects (see Ref. [89] for details). When the nitroxide was added to nitrogen-equilibrated beer at an initial concentration of 100 μM the kinetic curve started to plateau after c.10 h of the experiment, indicating that all the available reducing agents [H] were exhausted (Fig. 6.7). Experimental data were fitted to the simplest bimolecular reaction model assuming collisions of a nitroxide with just one reducing reagent in beer [6]. The fit residual, a difference between the experimental data points and the exponential fit (Fig. 6.7B), shows that this kinetic model describes the data exceptionally well. For fresh MGD beer, the rate constant of this bimolecular reaction was 54.8 ± 0.5/M/min and the concentration of the reducing agent was 61.7 ± 0.2 μM [6]. Based on the kinetic parameters and additional experiments, Kocherginsky and coworkers suggested that the observed chemical reaction of nitroxides in beer can be explained by a catalytic reduction with participation of SH groups of cysteines [6]. The fact that the kinetics of nitroxide reduction were affected by the state of proteins in beer (aggregation and denaturing upon boiling, and fragmentation upon pepsin treatment), and the accessibility of cysteines to nitroxides provided some strong arguments for this hypothesis [6].

Overall, the EPR detection method allows for accurate measurements of nitroxide reduction kinetics that could be analyzed to yield the effective concentration of the reducing agent(s), as well as the rate constant. The latter quantitative parameters could be used to characterize the redox power of beer and other beverages. Compared to spin-trapping reagents that are typically employed in high (40–60 mM) concentrations, the nitroxides could be applied at μM levels or even lower. The latter experiments could also be carried out using gas-permeable capillaries to remove oxygen to prevent reoxidation of the hydroxylamine form back to nitroxide—a problem that is commonly encountered in both nitroxide in DDPH reduction experiments. Finally, but not last, the experiments could be extended to other nitroxides with different standard redox potentials. The solubility of nitroxides in different solvents can be readily adjusted by chemical modifications of these very useful spin probe molecules.

6.4 MECHANISM OF PHOTO-INDUCED BEER OXIDATION BY TIME-RESOLVED ELECTRON PARAMAGNETIC RESONANCE

Free radical mediated oxidation processes in liquid food and beverages can also be studied by time-resolved EPR (TREPR). Experiments using TREPR are based on generating high non-equilibrium concentrations of free radicals by high-intensity short laser pulses, and detecting electronic spin transitions under continuous irradiation with microwave resonant frequency using a gated integrator (boxcar), or a transient digitizer if kinetic information is desired (e.g., see Ref. [90] for a review). For many systems, the photoexcitation and the resulting chemical reactions produce non-Boltzmann populations of the electronic spin states, a phenomenon known as chemically induced electron spin polarization (CIDEP) that leads to an additional enhancement of EPR signal intensity [91].

Forbes and coworkers were, perhaps, the first who took advantage of the high spectral resolution of TREPR to detect short-lived free radical species directly involved in the photodegradation of iso-α-acids (isohumulones)—the main bittering agents from hops in beer [10]. These authors employed laser flash photolysis at 308 nm to solutions of iso-α-acids in toluene/methylcyclohexane (1:1), and observed net emissive TREPR signals that were strongly spin polarized by the CIDEP triplet mechanism. The TREPR experiments identified an enolized β-triketone as the primary light-absorbing chromophore, and an uphill intramolecular triplet energy transfer process leading to Norrish type I α-cleavage at a second site, an α-hydroxycarbonyl [92]. Overall, the study identified the primary photophysics involved in the development of the undesirable "lightstruck" flavor of beer.

This work of Forbes and coworkers led to additional TREPR studies of photodegradation of isohumulones involved in the formation of the lightstruck flavor in beer [93]. Further details on the free radical intermediates were obtained by carefully designed and executed spin-trapping experiments that identified intermediate radicals from conventional CW EPR spectra of the spin adducts obtained [94–96]. This and other work that led to detailed understanding of the photo-induced free radical reaction that causes the lightstruck flavor in beer has been reviewed recently, together with other methods to study beer stability [14]. At the end, the authors concluded that as a result of these findings obtained mainly by EPR, "details about the reactions leading to lightstruck flavor formation may now be considered as fully unraveled" [14].

6.5 CONCLUSIONS AND OUTLOOK

In this chapter we reviewed the broad applications of EPR to study liquid food products and beverages. This field has grown tremendously, mainly over the last two decades. The EPR method is widely applicable to studying endogenous free radicals and paramagnetic metal ions, even though their role in food degradation and/or nutrition is not always understood. A broader integration of EPR results with other methods could be useful to achieve further progress in this area. The field of characterization of endogenous free radicals and metal ion complexes could also benefit from advanced EPR methods that offer superior resolution—from high field/high frequency EPR to double resonance methods such as HYSCORE (2D HYperfine Sub-level CORrElation spectroscopy) and ENDOR (Electron Nuclear DOuble Resonance). Currently, these methods are rarely applied to the characterization of food products. The problem of chemical heterogeneity of the endogenous species bearing unpaired electronic spins could be readily solved for liquid food products using modern chromatography methods.

Spin-trapping EPR approaches have proven to be very informative for the characterization of intermediate reactive free radical species, and could be widely applied to study the processes of food degradation and oxidation. These methods could be further combined with TREPR, which is uniquely suited to study photo-induced free radicals and radical pairs. Both approaches have been exceptionally successful in unraveling the free radical mechanism leading to the formation of a lightstruck flavor in beer, but could be used in a broader range of liquid food products.

Finally, but not last, EPR spin-traps and also nitroxide spin probes appear as very useful indicators of beer shelf life stability, as well as redox and antioxidant indicators. The nitroxide and DPPH EPR methods for antioxidant assessments reviewed in this chapter are now ready to compete with optical indicators as inexpensive, fully automated, table-top EPR spectrometers are now available from Bruker Biospin (Karlsruhe, Germany), Adani USA (Miami Lakes, FL), and Active Spectrum Inc. (Foster City, CA), among a few other suppliers.

ACKNOWLEDGMENT

The author would like to acknowledge the support of US DOE Contract DE-FG02-02ER15354 and specifically the development of least-squares simulation methods that were employed for the analysis of some of EPR spectra presented in this review.

REFERENCES

[1] P. Erdkamp, War and State Formation in the Roman Republic. A Companion to the Roman Army, Blackwell Publishing Ltd (2007), pp. 96–113.

[2] E.A. Decker, R.J. Elias, D.J. McClements (Eds.), Oxidation in Foods and Beverages and Antioxidant Applications, Woodhead Pub, Philadelphia, 2010.

[3] G.M. Rosen, B.E. Britigan, H.J. Halpern, S. Pou (Eds.), Free Radicals. Biology and Detection by Spinn Trapping, Oxford University Press, New York, 1999.

[4] G. Bačić, I. Spasojević, B. Šećerov, M. Mojović, Spin-trapping of oxygen free radicals in chemical and biological systems: new traps, radicals and possibilities, Spectrochim. Acta Part A Mol. Biomol. Spectrosc. 69 (5) (2008) 1354–1366.

[5] W.K.M. Chan, E.A. Decker, J.B. Lee, D.A. Butterfield, EPR spin-trapping studies of the hydroxyl radical scavenging activity of carnosine and related dipeptides, J. Agric. Food. Chem. 42 (7) (1994) 1407–1410.

[6] N.M. Kocherginsky, Y.Y. Kostetski, A.I. Smirnov, Use of nitroxide spin probes and electron paramagnetic resonance for assessing reducing power of beer. Role of SH groups, J. Agric. Food. Chem. 53 (4) (2005) 1052–1057.

[7] O.P. Sharma, T.K. Bhat, DPPH antioxidant assay revisited, Food. Chem. 113 (4) (2009) 1202–1205.

[8] J. Polak, M. Bartoszek, I. Stanimirova, A study of the antioxidant properties of beers using electron paramagnetic resonance, Food. Chem. 141 (3) (2013) 3042–3049.

[9] M.D.E. Forbes, L.E. Jarocha, S. Sim, V.F. Tarasov, Chapter one-time-resolved electron paramagnetic resonance spectroscopy: history, technique, and application to supramolecular and macromolecular chemistry, in: H.W. Ian, H.W. Nicholas, (Eds.), Advances in Physical Organic Chemistry, Academic Press (2013), pp. 1–83.

[10] C.S. Burns, A. Heyerick, D. De Keukeleire, M.D.E. Forbes, Mechanism for formation of the lightstruck flavor in beer revealed by time-resolved electron paramagnetic resonance, Chem. Eur. J. 7 (21) (2001) 4553–4561.

[11] M. Ono, M. Uchida, Inventors; Analytical Method for Evaluating Flavor Stability of Fermented Alcoholic Beverages Using Electron Spin Resonance. USA Patent 5,811,305. 1988 1988.

[12] D. Schmalbein, J. Jiang, A.H. Heiss, R.T. Weber, A. Kamlowski, Inventors; Method and Device for Analyzing the Stability of Fluid Foodstuffs by Method of Electron Spin Resonance Patent WO 2001036994, 2001.

[13] D. Barr, A. Heiss, A. Kamlowski, D. Maier, J. Erstling, H. Meling, Shelf life analysis of beer using an automated lag-time EPR system, Spectroscopy 16 (12) (2001) 16–19.

[14] K. Huvaere, M.L. Andersen, Beer and ESR spin trapping, in: V.R. Preedy (Ed.), Beer in Health and Disease Prevention, Academic Press, Amsterdam, 2009, pp. 1043–1053.

[15] D. Jehle, M.N. Lund, L.H. Ogendal, M.L. Andersen, Characterisation of a stable radical from dark roasted malt in wort and beer, Food. Chem. 125 (2) (2011) 380–387.

[16] G. Fritsch, C. Lopezt, J. Rodrigue, Generation and recombination of free-radicals in organic materials studied by electron-spin resonance, J. Magn. Reson. 16 (1) (1974) 48–55.

[17] A.I. Smirnov, R.L. Belford, R. Morse, Magnetic resonance imaging in a hands-on student experiment using an EPR spectrometer, Concep. Magn. Reson. 11 (5) (1999) 277–290.

[18] E.C. Pascual, B.A. Goodman, C. Yeretzian, Characterization of free radicals in soluble coffee by electron paramagnetic resonance spectroscopy, J. Agric. Food. Chem. 50 (21) (2002) 6114–6122.

[19] B.A. Goodman, E.C. Pascual, C. Yeretzian, Real time monitoring of free radical processes during the roasting of coffee beans using electron paramagnetic resonance spectroscopy, Food. Chem. 125 (2011) 248–254.

[20] G.J. Troup, L. Navarini, F.S. Liverani, S.C. Drew, Stable radical content and anti-radical activity of roasted Arabica coffee: from in-tact bean to coffee brew, PLoS One 10 (4) (2015) e0122834.

[21] J. Gonis, D.G. Hewitt, G. Troup, D.R. Hutton, C.R. Hunter, The chemical origin of free radicals in coffee and other beverages, Rad. Res. 23 (1995) 393–399.

[22] G.J. Troup, D.R. Hutton, D.G. Hewitt, C.R. Hunter, Free-radicals in red wine, but not in white, Free. Radic. Res. 20 (1) (1994) 63–68.

[23] J.A. Kennedy, G.J. Troup, J.R. Pilbrow, D.R. Hutton, D. Hewitt, C.R. Hunter, et al., Development of seed polyphenols in berries from *Vitis vinifera* L. cv. Shiraz, Aust. J. Grape Wine Res. 6 (3) (2000) 244–254.

[24] D.K. Das, M. Sato, P.S. Ray, G. Maulik, R.M. Engelman, A.A.E. Bertelli, et al., Cardioprotection of red wine: role of polyphenolic antioxidants, Drugs Exp. Clin. Res. 25 (2–3) (1999) 115–120.

[25] M.A. Morsy, M.M. Khaled, Novel EPR characterization of the antioxidant activity of tea leaves, Spectrochim. Acta Part A Mol. Biomol. Spectrosc. 58 (6) (2002) 1271–1277.

[26] M. Polovka, V. Brezova, A. Stasko, Antioxidant properties of tea investigated by EPR spectroscopy, Biophys. Chem. 106 (1) (2003) 39–56.

[27] M. Polovka, EPR spectroscopy: a tool to characterize stability and antioxidant properties of foods, J. Food Nutr. Res. 45 (1) (2006) 1–11.

[28] R. Biyik, R. Tapramaz, An EPR study on tea: identification of paramagnetic species, effect of heat and sweeteners, Spectrochim. Acta Part A Mol. Biomol. Spectrosc. 74 (3) (2009) 767–770.

[29] J. Emsley, Nature's Building Blocks: An A–Z Guide to the Elements, Oxford University Press, Oxford, UK, 2011.

[30] A. Abragam, B. Bleaney, Electron Paramagnetic Resonance of Transition Ions, Oxford University Press, Oxford, UK, 2012.

[31] F.A. Taiwo, Electron paramagnetic resonance spectroscopic studies of iron and copper proteins, Spectrosc. Int. J. 17 (1) (2003) 53–63.

[32] B. Halliwell, J.M.C. Gutteridge, The importance of free-radicals and catalytic metal-ions in human-diseases, Mol. Aspects. Med. 8 (2) (1985) 89–193.

[33] H. Kaneda, Y. Kano, S. Koshino, H. Ohyanishiguchi, Behavior and role of iron ions in beer deterioration, J. Agric. Food. Chem. 40 (11) (1992) 2102–2107.

[34] A.I. Smirnov, O.E. Yakimchenko, H.A. Golovina, S.K. Bekova, Y.S. Lebedev, EPR imaging with natural spin probes, J. Magn. Reson. 91 (2) (1991) 386–391.

[35] G.R. Buettner, Spin trapping-electron-spin-resonance parameters of spin adducts, Free Radic. Biol. Med. 3 (4) (1987) 259–303.

[36] V. Khramtsov, L.J. Berliner, T.L. Clanton, NMR spin trapping: detection of free radical reactions using a phosphorus-containing nitrone spin trap, Magn. Reson. Med. 42 (2) (1999) 228–234.

[37] E.G. Janzen, P.H. Krygsman, D.A. Lindsay, D.L. Haire, Detection of alkyl, alkoxyl, and alkyperoxyl radicals from the thermolysis of azobis(isobutyronitrile) by ESR/spin trapping. Evidence for double spin adducts from liquid-phase chromatography and mass spectroscopy, J. Am. Chem. Soc. 112 (23) (1990) 8279–8284.

[38] R.P. Mason, Using anti-5,5-dimethyl-1-pyrroline N-oxide (anti-DMPO) to detect protein radicals in time and space with immuno-spin trapping, Free Radic. Biol. Med. 36 (10) (2004) 1214–1223.

[39] M. Conte, H. Miyamura, S. Kobayashi, V. Chechik, Enhanced acyl radical formation in the Au nanoparticle-catalysed aldehyde oxidation, Chem. Commun. 46 (1) (2010) 145–147.

[40] T.I. Smirnova, A.I. Smirnov, R.B. Clarkson, R.L. Belford, Y. Kotake, E.G. Janzen, High-frequency (95 GHz) EPR spectroscopy to characterize spin adducts, J. Phys. Chem. B 101 (19) (1997) 3877–3885.

[41] H. Kaneda, Y. Kano, T. Osawa, N. Ramarathnam, S. Kawakishi, K. Kamada, Detection of free-radicals in beer oxidation, J. Food. Sci. 53 (3) (1988) 885–888.

[42] H. Kaneda, Y. Kano, T. Osawa, S. Kawakishi, S. Koshino, Free-radical reactions in beer during pasteurization, Int. J. Food Sci. Technol. 29 (2) (1994) 195–200.

[43] M. Uchida, M. Ono, Improvement for oxidative flavor stability of beer – role of OH-radical in beer oxidation, J. Am. Soc. Brew. Chem. 54 (4) (1996) 198–204.

[44] M. Uchida, S. Suga, M. Ono, Improvement for oxidative flavor stability of beer – rapid prediction method for beer flavor stability by electron spin resonance spectroscopy, J. Am. Soc. Brew. Chem. 54 (4) (1996) 205–211.

[45] M.L. Andersen, L.H. Skibsted, Electron spin resonance spin trapping identification of radicals formed during aerobic forced aging of beer, J. Agric. Food. Chem. 46 (4) (1998) 1272–1275.

[46] M.L. Andersen, H. Outtrup, L.H. Skibsted, Potential antioxidants in beer assessed by ESR spin trapping, J. Agric. Food. Chem. 48 (8) (2000) 3106–3111.

[47] V. Brezova, M. Polovka, A. Stasko, The influence of additives on beer stability investigated by EPR spectroscopy, Spectrochim. Acta Part A Mol. Biomol. Spectrosc. 58 (6) (2002) 1279–1291.

[48] N.M. Kocherginsky, Y.Y. Kostetski, A.I. Smirnov, Antioxidant pool in beer and kinetics of EPR spin-trapping, J. Agric. Food. Chem. 53 (17) (2005) 6870–6876.

[49] B.E. Britigan, S. Pou, G.M. Rosen, D.M. Lilleg, G.R. Buettner, Hydroxyl radical is not a product of the reaction of xanthine-oxidase and xanthine – the confounding problem of adventitious iron bound to xanthine-oxidase, J. Biol. Chem. 265 (29) (1990) 17533–17538.

[50] A.I. Smirnov, R.L. Belford, Rapid quantitation from inhomogeneously broadened EPR spectra by a fast convolution algorithm, J. Magn. Reson. Ser. A 113 (1) (1995) 65–73.

[51] A.N. Saprin, L.H. Piette, Spin trapping and its application in study of lipid peroxidation and free-radical production with liver-microsomes, Arch. Biochem. Biophys. 180 (2) (1977) 480–492.

[52] G.V. Buxton, C.L. Greenstock, W.P. Helman, A.B. Ross, Critical-review of rate constants for reactions of hydrated electrons, hydrogen-atoms and hydroxyl radicals (.Oh/.O-) in aqueous-solution, J. Phys. Chem. Ref. Data 17 (2) (1988) 513–886.

[53] D. Barr, B. Baker, S. Bosben, T. Bradshaw, E. Converse, A. Daar, et al., Standard method for measurement of oxidative resistance of beer by electron paramagnetic resonance, J. Am. Soc. Brew. Chem. 66 (4) (2008) 259–260.

[54] D. Barr, S. Bobsen, T. Bradshaw, C. Giarratano, H. Hight, K. Holsopple, et al., Method for measurement of resistance of oxidation in beer by electron paramagnetic resonance, J. Am. Soc. Brew. Chem. 65 (4) (2007) 244–245.

[55] J. Liu, J.J. Dong, Q. Li, J. Chen, G.X. Gu, Investigation of new indexes to evaluate aging of bottled lager beer, J. Am. Soc. Brew. Chem. 66 (3) (2008) 167–173.

[56] C. Laane, G. de Roo, E. van den Ban, M.W. Sjauw-En-Wa, M.G. Duyvis, W.A. Hagen, et al., The role of riboflavin in beer flavour instability: EPR studies and the application of flavin binding proteins, J. Inst. Brew. 105 (6) (1999) 392–397.

[57] A.M. Frederiksen, R.M. Festersen, M.L. Andersen, Oxidative reactions during early stages of beer brewing studied by electron spin resonance and spin trapping, J. Agric. Food. Chem. 56 (18) (2008) 8514–8520.

[58] M.L. Andersen, L.H. Skibsted, Modification of the levels of polyphenols in wort and beer by addition of hexamethylenetetramine or sulfite during mashing, J. Agric. Food. Chem. 49 (11) (2001) 5232–5237.

[59] A.F. Suarez, T. Kunz, N.C. Rodriguez, J. MacKinlay, P. Hughes, F.J. Methner, Impact of colour adjustment on flavour stability of pale lager beers with a range of distinct colouring agents, Food. Chem. 125 (3) (2011) 850–859.

[60] E. Jeney-Nagymate, P. Fodor, Examination of the effect of vitamin E and C addition on the beer's ESR lag time parameter, J. Inst. Brew. 113 (1) (2007) 28–33.

[61] C.M. Fraser, J.D. Gocayne, O. White, M.D. Adams, R.A. Clayton, R.D. Fleischmann, et al., The minimal gene complement of mycoplasma-genitalium, Science 270 (5235) (1995) 397–403.

[62] M.K. Thomsen, D. Kristensen, L.H. Skibsted, Electron spin resonance spectroscopy for determination of the oxidative stability of food lipids, J. Am. Oil Chem. Soc. 77 (7) (2000) 725–730.

[63] J. Velasco, M.L. Andersen, L.H. Skibsted, Evaluation of oxidative stability of vegetable oils by monitoring the tendency to radical formation. A comparison of electron spin resonance spectroscopy with the Rancimat method and differential scanning calorimetry, Food. Chem. 85 (4) (2004) 623–632.

[64] A. Stagko, M. Liptakova, F. Malik, V. Misik, Free radical scavenging activities of white and red wines: an EPR spin trapping study, Appl. Magn. Reson. 22 (1) (2002) 101–113.

[65] R.J. Elias, A.L. Waterhouse, Controlling the fenton reaction in wine, J. Agric. Food. Chem. 58 (3) (2010) 1699–1707.

[66] R.J. Elias, M.L. Andersen, L.H. Skibsted, A.L. Waterhouse, Identification of free radical intermediates in oxidized wine using electron paramagnetic resonance spin trapping, J. Agric. Food. Chem. 57 (10) (2009) 4359–4365.

[67] R.J. Elias, M.L. Andersen, L.H. Skibsted, A.L. Waterhouse, Key factors affecting radical formation in wine studied by spin trapping and EPR spectroscopy, Am. J. Enol. Vitic. 60 (4) (2009) 471–476.

[68] S.B. Nimse, D. Pal, Free radicals, natural antioxidants, and their reaction mechanisms, RSC Adv. 5 (35) (2015) 27986–28006.

[69] N.E.C. de Almeida, P. Homem-de-Mello, D. De Keukeleire, D.R. Cardoso, Reactivity of beer bitter acids toward the 1-hydroxyethyl radical as probed by spin-trapping electron paramagnetic resonance (EPR) and electrospray ionization-tandem mass spectrometry (ESI-MS/MS), J. Agric. Food. Chem. 59 (8) (2011) 4183–4191.

[70] N.E.C. de Almeida, E.S.P. do Nascimento, D.R. Cardoso, On the reaction of lupulones, hops beta-acids, with 1-hydroxyethyl radical, J. Agric. Food. Chem. 60 (42) (2012) 10649–10656.

[71] G.C. Yen, H.Y. Chen, Antioxidant activity of various tea extracts in relation to their antimutagenicit, J. Agric. Food. Chem. 43 (1) (1995) 27–32.

[72] M.S. Blois, Antioxidant determinations by the use of a stable free radical, Nature. 181 (4617) (1958) 1199–1200.

[73] K. Mishra, H. Ojha, N.K. Chaudhury, Estimation of antiradical properties of antioxidants using DPPH center dot assay: a critical review and results, Food. Chem. 130 (4) (2012) 1036–1043.

[74] V. Brezová, A. Šlebodová, A. Staško, Coffee as a source of antioxidants: an EPR study, Food. Chem. 114 (3) (2009) 859–868.

[75] J. Deng, W. Cheng, G. Yang, A novel antioxidant activity index (AAU) for natural products using the DPPH assay, Food. Chem. 125 (4) (2011) 1430–1435.

[76] M. Jurkova, T. Horak, D. Haskova, J. Culik, P. Cejka, V. Kellner, Control of antioxidant beer activity by the mashing process, J. Inst. Brew. 118 (2) (2012) 230–235.

[77] J.A. Larrauri, C. Sanchez-Moreno, P. Ruperez, F. Saura-Calixto, Free radical scavenging capacity in the aging of selected red Spanish wines, J. Agric. Food. Chem. 47 (4) (1999) 1603–1606.

[78] M. Bartosek, J. Polak, An electron paramagnetic resonance study of antioxidant properties of alcoholic beverages, Food. Chem. 132 (4) (2012) 2089–2093.

[79] M. Bartosek, J. Polak, A comparison of antioxidative capacities of fruit juices, drinks and nectars, as determined by EPR and UV-vis spectroscopies, Spectrochim. Acta Part A Mol. Biomol. Spectrosc. 153 (2016) 546–549.

[80] M. Zalibera, A. Stasko, A. Slebodova, V. Jancovicova, T. Cermakova, V. Brezova, Antioxidant and radical-scavenging activities of Slovak honeys – an electron paramagnetic resonance study, Food. Chem. 110 (2) (2008) 512–521.

[81] H. Kaneda, N. Kobayashi, S. Furusho, H. Sahara, S. Koshino, Chemical evaluation of beer flavor stability, MBAA Tech. Q. 32 (1995) 76–80.

[82] G.I. Likhtenshtein, J. Yamauchi, S. Nakatsuji, A.I. Smirnov, R. Tamura, Nitroxides: Applications in Chemistry, Biomedicine, and Materials Science, Wiley-VCH, Weinheim, Germany, 2008.

[83] T.I. Smirnova, M.A. Voinov, A.I. Smirnov, Spin Probes and Spin Labels. Encyclopedia of Analytical Chemistry, John Wiley & Sons, Ltd (2009).

[84] L.J. Berliner (Ed.), Spin Labeling: Theory and Applications, Academic Press, New York, NY, 1976.

[85] G.I. Likhtenshtein, J. Yamauchi, Si Nakatsuji, A.I. Smirnov, R. Tamura, Nitroxide Redox Probes and Traps, Nitron Spin Traps, Nitroxides: Wiley-VCH Verlag GmbH & Co. KGaA (2008), pp. 239–268.

[86] G.I. Shchukin, I.A. Grigor'ev, Oxidation-reduction properties of nitroxides, in: L.B. Volodarsky (Ed.), Imidazoline Nitroxides, CRC Press, Boca Raton, FL, 1988, pp. 171–214.

[87] N. Kocherginsky, H.M. Swartz, Nitroxide Spin Labels: Reactions in Biology and Chemistry, CRC Press, Boca Raton, FL, 1995.

[88] A.I. Smirnov, R.B. Clarkson, R.L. Belford, EPR linewidth (T-2) method to measure oxygen permeability of phospholipid bilayers and its use to study the effect of low ethanol concentrations, J. Magn. Reson. Ser. B 111 (2) (1996) 149–157.

[89] P.D. Morse, A.I. Smirnov, Simultaneous ESR measurements of the kinetics of oxygen consumption and spin label reduction by mammalian cells, Magn. Reson. Chem. 33 (1995) S46–S52.

[90] M.D.E. Forbes, Time-resolved (CW) electron paramagnetic resonance spectroscopy: an overview of the technique and its use in organic photochemistry, Photochem. Photobiol. 65 (1) (1997) 73–81.

[91] G.H. Goudsmit, H. Paul, A.I. Shushin, Electron-spin polarization in radical triplet pairs – size and dependence on diffusion, J. Phys. Chem. 97 (50) (1993) 13243–13249.

[92] C.S. Burns, A. Heyerick, D. De Keukeleire, M.D.E. Forbes, Mechanism for formation of the lightstruck flavor in beer revealed by time-resolved electron paramagnetic resonance, Chem. Eur. J. 7 (21) (2001) 4553–4561.

[93] A. Heyerick, K. Huvaere, D. De Keukeleire, M.D.E. Forbes, Fate of flavins in sensitized photodegradation of isohumulones and reduced derivatives: studies on formation of radicals via EPR combined with detailed product analyses, Photochem. Photobiol. Sci. 4 (5) (2005) 412–419.

[94] K. Huvaere, M.L. Andersen, K. Olsen, L.H. Skibsted, A. Heyerick, D. De Keukeleire, Radicaloid-type oxidative decomposition of beer bittering agents revealed, Chem. Eur. J. 9 (19) (2003) 4693–4699.

[95] K. Huvaere, M.L. Andersen, L.H. Skibsted, A. Heyerick, D. De Keukeleire, Photooxidative degradation of beer bittering principles: a key step on the route to lightstruck flavor formation in beer, J. Agric. Food. Chem. 53 (5) (2005) 1489–1494.

[96] K. Huvaere, M.L. Andersen, M. Storme, J. Van Bocxlaer, L.H. Skibsted, D. De Keukeleire, Flavin-induced photodecomposition of sulfur-containing amino acids is decisive in the formation of beer lightstruck flavor, Photochem. Photobiol. Sci. 5 (10) (2006) 961–969.

CHAPTER 7

ESR of Irradiated Drugs and Excipients for Drug Control and Safety

S. Iravani
Isfahan University of Medical Sciences, Isfahan, Iran

Contents

7.1 INTRODUCTION

In addition to heat and gas exposure sterilization, ionizing radiation is gaining attention for the sterilization of pharmaceuticals and medicinal products. Gamma radiation processes can be used to achieve safe and effective sterilization of pharmaceutical products, disposable medical devices, and other medicinal products (especially in the case of thermolabile products). It does not produce chemical contamination, and there are no chemical residues after the sterilization process has been completed. In this field, the detection and dosimetry of pharmaceutical radio-sterilization is an important concern for numerous pharmaceutical regulatory agencies from around the world.

Electron Spin Resonance in Food Science.
111

Electron spin resonance (ESR) can be used as an effective and powerful technique for studying chemical species or materials which have one or more unpaired electrons. In fact, a huge literature has been published on the basic physical concepts of ESR and its applications [1–7]. Electron spin resonance spectrometry can be used as a suitable analytical method for studying chemical species and materials. Nowadays, ESR has found important applications as a dosimetric tool including alanine dosimetry, individual doses from retrospective dosimetry after radiation accidents, ESR dating, identification of irradiated foodstuffs, and paramagnetic center concentration in materials [8,9]. For drugs irradiated in a solid and relatively dry state, the induced and trapped radicals and ions can be detected by ESR methods [8,9]. A lot of drugs and excipients can be irradiated in the solid state; generally, they show no ESR signal in the unirradiated sample, which is consequently very easy to detect. In the case of natural products used in some drugs, the unirradiated sample may also present a single line, as is very often found in vegetal products probably due to a quinone radical, but this single line is easy to distinguish from the complex signal induced in irradiated drugs [8,9]. The ESR analysis makes nondestructive, reproducible, irradiation specific, and time-efficient results available [10]. In this chapter, some important examples of the ESR spectrometry analysis of irradiated drugs and expedients are mentioned.

7.2 ELECTRON SPIN RESONANCE OF IRRADIATED DRUGS AND EXCIPIENTS

7.2.1 Antibiotics

The ESR technique was used to detect antibiotics irradiated for sterilization purposes [11]. For instance, 13 different cephalosporins were given a sterilization dose of 25 kGy and then studied by ESR spectrometry. The two basic requirements of specificity and stability are met by the ESR technique when used to detect irradiated cephalosporins [11]. In another study, the ESR spectrum of microspheres containing vancomycin irradiated at 25 kGy was reported. As a result, the ESR spectrum observed in irradiated microspheres consisted of several lines, indicating the presence of more than one type of free radical species induced by γ-irradiation. The signal intensity in the control sample (even though magnified 40 times) was negligible with respect to the irradiated sample, the ratio between the two signal intensities was actually about 400 [12]. Basly et al. [13] used an ESR technique for identification and quantification purposes in irradiated antibiotics including latamoxef (an oxacephem antibiotic) and ceftriaxone (a cephalosporin). As a result, the limit of detection and limit of discrimination were (0.5 kGy, 1.5 kGy) and (1.5 kGy, 5 kGy), for latamoxef and ceftriaxone, respectively. The limits of detection of free radicals after irradiation at 25 kGy were 140 days for latamoxef, and 115 days for ceftriaxone. In addition, ESR spectroscopy was used for analysis radio-sterilization of cefoperazone (a cephalosporin). As a result, while the ESR spectra of an unirradiated sample demonstrated no intensity, a signal, dependent on the irradiation dose, was found exclusively in irradiated samples [14]. Gibella et al. [15] studied the ESR spectroscopy of five cephalosporins and penicillin antibiotics including anhydrous ampicilline acid, amoxicilline acid trihydrate, cefuroxime sodium salt, cloxacilline sodium salt monohydrate, and ceftazidime pentahydrate. Consequently, the ESR analysis demonstrated the influence of irradiation and storage parameters on the nature and concentration of the free radicals trapped. By using this analytical method, it seems that researchers can distinguish the irradiated samples from the unirradiated ones. In another study, the concentrations and properties of free radicals in piperacillin, ampicillin, and crystalline penicillin after γ-irradiation were determined using EPR analysis (using an X-band spectrometer, at 9.3 GHz). Gamma irradiation was performed at a dose of 25 kGy. After γ-irradiation, complex EPR lines were recorded, confirming the presence of a large number of free radicals formed during the irradiation. It was reported that for all tested antibiotics, concentrations of free radicals and parameters of ESR spectra changed with storage time [16].

In order to investigate the concentration and dynamics of free radicals generated in antibiotics due to thermal sterilization, ESR studies of two antibiotics including cefaclor (a semisynthetic antibiotic chemically related to penicillin) and clarithromycin (a macrolide antibiotic) were performed. For cefaclor, three combinations of temperature and heating time were applied: 160°C, 170°C, and 180°C for 120, 60, and 30 min, respectively. Clarithromycin was heated at 160°C for 120 min, and the ESR line shape was investigated versus microwave power ranging from 2.2 to 70 mW. Electron spin–spin relaxation time was estimated from the ESR line shape analysis. It was concluded that while the relaxation features of the centers were similar for both compounds, they were not considerably affected by the sterilization procedure. Moreover, it was reported that the concentration of the paramagnetic centers was, however, strongly dependent on the details of sterilization [17]. In another study, ESR spectroscopy was used to study the degradation gamma-damage in a microcrystalline powder from azithromycin (a macrolide) before and after exposure to a 5 up to 25 KGy absorbed dose. ESR measurements proved various stable paramagnetic species after irradiation and relative yielding of free radicals depended on the absorbed gamma dose. As a result, from the analysis of ESR signal dependence on the absorbed dose, it seems that γ-irradiation caused increases in the amount of radicals in samples [18]. Furthermore, the effect of γ-irradiation from a ^{60}Co source with a 2MeV electron beam was studied on two fluoroquinolone antibiotics (norfloxacin and gatifloxacin) in the solid state. As a result, ESR analysis demonstrated the number of free radical species formed and their population [19].

7.2.2 Aspirin

In one study, Cozar et al. [20] analyzed the ESR spectrum of the radiation damage in a powder of aspirin (2-acetoxybenzoic acid), one of the most used anti-inflammatory drugs. As a result, three types of radicals occur by γ-irradiation of aspirin at room temperature. Two of them were the result of hydrogen abstraction, while the third was produced by hydrogen addition at one of the carbon atoms of the ring.

7.2.3 Nitroimidazoles

In one study, Duroux et al. [21] examined the time stability at ambient conditions and higher temperatures of free radicals produced after γ-irradiation of three nitroimidazoles: metronidazole, ornidazole, and ternidazole. As a result, γ-irradiation of metronidazole and ornidazole produced free radicals which were detectable by ESR and appeared relatively stable. The bi-exponential model used to fit the decay curves with time at ambient conditions and high temperatures was acceptable.

7.2.4 Anti-Emetic Drugs

The behaviors of the γ-radiation-induced radicals in some anti-emetic drugs, including metoclopramide and odansetron, using ESR spectroscopy was investigated in order to characterize the specific features of these radicals depending on the absorbed dose. Consequently, ESR measurements indicated that both of them contained various stable paramagnetic species after irradiation, and the relative yield of free radicals depended on the absorbed dose. The ESR spectra of metoclopramide samples exhibited a single signal without hyperfine structure, centered on $g = 2.0047$ and a linewidth of 20 G, assumed to be due to the presence of one radical centered on carbon atoms. The fact that by increasing the irradiation dose, in the sample of odansetron, the different ESR lines did not vary in the same way, proved the presence in the irradiated sample, simultaneously, of radicals with different magnetic parameters [22].

7.2.5 Diclofenac Sodium

One of the most important problems of radio-sterilization is the production of new radiolytic products during the irradiation process. Therefore, the principal problem in radio-sterilization is to determine and to characterize these physical and chemical changes originating from high-energy radiation. In one study, γ-irradiation was applied as a sterilization method for diclofenac sodium and the raw materials dimyristoylphosphatidylcholine, surfactant I [polyglyceryl-3-cethyl ether (SUR I)], dicetyl phosphate, and cholesterol were used to prepare the system. The raw materials were irradiated with different radiation doses (5, 10, 25, and 50 kGy) and ESR characteristics were studied under normal (25°C, 60% relative humidity) and accelerated (40°C, 75% relative humidity) stability test conditions. As a result, a model based on three radical species was found to explain the experimental results obtained for γ-irradiated diclofenac sodium well. The radiation yield of solid diclofenac sodium was not high even at a dose of 50 kGy, and the radiolytical intermediates produced in diclofenac sodium decayed rapidly at room temperature. The irradiated sample stored under accelerated stability conditions for 12 h exhibited no ESR signal. This means that solid diclofenac sodium and drugs containing diclofenac sodium as an active ingredient could be safely sterilized by irradiation, as the radicals decay in a day when it is stored under stable conditions and it has quenched rapidly [23].

7.2.6 Herbal Products

In herbal products, irradiation produces free radicals in cellulose and crystalline sugars which could serve as irradiation detection markers in ESR spectroscopy studies. For instance, in an interesting study, radiation-induced free radicals were studied in the seeds of alfalfa (*Medicago sativa*) and broccoli (*Brassica oleracea*) irradiated at 0, 2, 4, and 6 kGy. As a result, freeze-dried and alcoholic extracted samples lacked radiation-induced ESR spectral characteristics. However, a sample treatment with 5% nitric acid was found to be the most appropriate to obtain clear evidence of irradiation. Results showed that freeze-drying and alcoholic extraction could be used to remove moisture, but the seed samples did not show a response to ESR-based detection. It seems that nitric acid extraction for irradiated alfalfa and broccoli sprout seeds proved to be a suitable treatment for ESR analysis. The spectrum with two side peaks of $g_1 = 2.0245$ and $g_2 = 1.9955$ was associated with typical radiation-induced ESR signals in alfalfa seeds. The irradiated broccoli seed samples showed typical ESR spectra with a doublet centered at $g = 2.0$ giving a clear indication of the hydroxyalkyl radical in the crystalline carbohydrates [24]. In another study, an ESR spectrometry method for five seeds including evening primrose seed, safflower seed, rape seed, sunflower seed, and flax seed were analyzed. Samples were irradiated at 1~10 kGy using a gamma-ray irradiator. As a result, the ESR signal (single-line) intensity of irradiated foods was higher than nonirradiated foods. The hydrocarbons 1,7-hexadecadiene ($C_{16:2}$) and 8-heptadecene ($C_{17:1}$) from oleic acid were detected only in the irradiated samples before and after treatment at doses of ≥ 1 kGy, but they were not detected in nonirradiated samples before and after treatment. These two hydrocarbons could be used as markers to identify irradiated safflower, rape, sunflower, and flax seeds. Then, the hydrocarbons 1,7,10-hexadeca-triene ($C_{16:3}$) and 6,9-heptadecadiene ($C_{17:2}$) from linoleic acid were detected in the evening primrose, safflower, and sunflower seeds [25].

Ukai et al. [26] analyzed ESR signals of dry vegetables before and after irradiation. As a result, before irradiation, the ESR signal of dry vegetables consisted of three components: a singlet at $g = 2.0030$, the sextet signals from Mn^{2+} ions, and a singlet from Fe^{3+}. The first originated from a carbon centered organic free radical. The second is attributable to the sextet signal with hyperfine interactions of Mn^{2+} ions centered at $g = 2.0020$. The third is a singlet at $g = 4.0030$ due to Fe^{3+}. After the γ-ray irradiation, a new pair of signals, or twin peaks, appeared in the ESR spectrum of dry vegetables. The intensity of the organic free radical at $g = 2.0030$ of the irradiated dry vegetables increased linearly with radiation doses. The progressive saturation behavior of the dry vegetables indicated a unique saturation, and the signals obeyed various relaxation processes. In another study, Nakamura et al. [27] revealed the presence of four radical species in γ-ray-irradiated ginseng (*Agaliaceae*). Before irradiation, the representative ESR spectrum of ginseng is composed of a sextet centered at $g = 2.0$, a sharp singlet at the same g-value, and a singlet at about $g = 4.0$. The first one is attributable to a hyperfine signal of Mn^{2+} ions (hyperfine constant: 7.4 mT). The second one is due to an organic free radical. The third one originates from Fe^{3+}. Upon γ-ray irradiation, a new ESR (the fourth) signal was detectable in the vicinity of the $g = 2.0$ region. The progressive saturation behaviors of the ESR signals at various microwave power levels were indicative of different relaxation times for those radicals. Anisotropic ESR spectra were detected by angular rotation of the sample tube. It is because of the existence of anisotropic microcrystalline in ginseng powder sample. Yamaoki et al. [28] analyzed unpaired electron components in royal jelly using ESR. As a result, the ESR spectrum of royal jelly had natural signals derived from transition metals, including Fe^{3+} and Cu^{2+}, and a signal line near $g = 2.00$. After irradiation, a new splitting asymmetric spectrum with an overall spectrum width of ca. 10 mT at $g = 2.004$ was observed. The intensities of the signals at $g = 2.004$ increased in proportion to the absorbed dose in samples under different storage conditions: fresh frozen royal jelly, and dried royal jelly powder at room temperature. The signal intensity of the fresh frozen sample was stable after irradiation. One year after 10 kGy irradiation of the dried powder, the signal intensity was sevenfold greater than before irradiation, although the intensity continued to steadily decrease with time. This stable radiation-induced radical component was derived from the poorly soluble constituent of royal jelly. Furthermore, in one study, the ESR spectral property of irradiated peony root was studied during the development of a system for detection of irradiated medicinal plants containing crystalline sugars. The plant had a weak ESR signal near $g = 2.005$ before electron beam-irradiation. After 10 kGy irradiation, the line shape became broader and cellulose-like signal (a pair of lines at 6 mT across the center signal near $g = 2.005$) and sugar-like spectrum (a broader line with more than 6 mT of overall spectrum width) was observed. The ESR intensity decreased considerably at 30 days postirradiation and was stable thereafter. The stable ESR line shape of irradiated peony root resembled that of irradiated sucrose, and the intensity 30 days postirradiation increased linearly as a function of sucrose content [29].

7.2.7 Drug Delivery Systems and Active Packaging Films

The ESR spectroscopy analysis of some drug delivery systems and active packaging films has been studied. For instance, in one study, irradiated active packaging films based on chitosan—fish gelatin containing coumarin (a fragrant organic chemical compound used as an antioxidant) were studied. After the drying process, the films were irradiated at 40 and 60 kGy using an electron beam accelerator. The effect of irradiation on the film properties, as well as the coumarin release mechanism, was studied and compared with the control. Consequently, ESR revealed free radical formation during irradiation in films containing coumarin. Antioxidant addition and/or irradiation treatment at a dose of 60 kGy resulted in a shift of amide A and amide B peaks. Moreover, a shift of the amide II band was only observed for the control film at the same dose. At 60 kGy, the tensile strength of only the control films increased significantly [30]. In another study, Faucitano et al. [31] investigated the primary and secondary free radical intermediates in the gamma radiolysis of poly(D,L-lactide-*co*-glycolide) and poly(D,L-lactide-*co*-glycolide) microspheres loaded with clonazepam using matrix ESR spectroscopy in the temperature range 77–298 K. Consequently, it was found that the drug–polymer interaction lead to significant deviations in the G (radicals) from the additivity-law. Furthermore, a radio-stabilization effect observed by a decrease of the overall radical yield in the mixed system was accompanied by an increase of the drug radical yield. Therefore, a transfer of the radiation damage from the polymer matrix to the drug had occurred which was partly attributed to the radical scavenging action of the nitro group, leading to relatively stable nitroxyl adducts [31].

7.2.8 Methylxanthine Derivatives

The radio chemical stability of three derivatives of methylxanthine including caffeine, theophylline, and theobromine in solid phase was studied. These drugs in solid phase were irradiated with ionizing radiation produced by a beam of electrons of energy 9.96 MeV in doses of 25, 50, 100, 200, and 400 kGy. The ESR analysis was carried out for all nonirradiated and irradiated samples in standard sample tubes. Findings from this study demonstrated that the methylxanthine derivatives were resistant to ionizing irradiation in the doses usually used for sterilization (<50 kGy), and they were relatively radio-chemically stable and could be sterilized by irradiation [32].

7.2.9 Cimetidine and Famotidine

Sayin investigated the magnetic properties of γ-irradiated single crystal cimetidine between temperatures of 123 and 418 K, and between microwave powers of 0.01 and 150 mW by ESR analysis techniques. Magnetic field orientation in each of the three perpendicular axes, microwave power, and temperature dependence of the EPR spectra led to more than one radical being produced by gamma irradiation in the host crystal.

The distinctive radical was attributed to a ring type radical, but the other radicals could not be identified because of superimposition [33]. Moreover, in another study, free radical formation in thermally sterilized famotidine was studied using X-band (9.3 GHz) ESR spectroscopy. Sterilization was done at temperatures of 160°C (120 min), 170°C (60 min), and 180°C (30 min). As a result, it was found that the optimal temperature of thermal sterilization for famotidine was 170°C (60 min), and in this condition the lowest free radical concentration was found. The highest free radical concentration was measured in famotidine heated at 160°C (120 min) [34].

7.2.10 Ascorbic Acid and Sodium Ascorbate

Basly et al. [35] reported ESR analysis of γ-irradiated ascorbic acid and sodium ascorbate. As a result, while the ESR spectra of nonirradiated samples presented no intensity, a signal, dependent on the irradiation dose, was found in irradiated samples.

7.2.11 Nitrofurans

In one study, experimental data on ESR dosimetry of irradiated nitrofurans including nitrofurantoin, nifuroxazide, nifurzide, and nifurtoinol was reported. As a result, the ESR spectrum of nonirradiated samples showed no signal, and a signal which was dependent on the irradiation dose was observed with irradiated samples [36].

7.2.12 Neurological and Antihypertensive Drugs

The ESR of some γ-irradiated neurological and antihypertensive drugs (such as sodium valproate, pentoxifylline, selegiline, pergolid mesylate, lisinopril, diltiazem, etc.) was investigated. It was found that γ-irradiation produced some very stable alkyl- and amine-type free radicals in these drugs. Analysis of the samples demonstrated no ESR signal without irradiation [37]. In another study, exposure of dry powder forms of the drugs, including nitrendipine, nifedipine, felodipine, and nimodipine, to γ-irradiation resulted in the formation of free radicals detected by ESR. The four structurally related drugs showed qualitatively identical ESR spectral features in terms of g-values (the qualitative descriptive parameter). Because of the high stability of the radicals produced, it seems that administration of these irradiated drugs may present patients with quantities of free radicals and the possibility of secondary cell damage [38].

7.3 CONCLUSIONS

In general, high-energy ionizing radiation, such as γ-irradiation, can be used for the sterilization of medical devices and pharmaceutical products, and improvement in the hygienic quality of foods. After radiation sterilization, it is always necessary to reveal that any products or ingredients formed during the irradiation are not harmful or hazardous. ESR can be used to study radiolysis mechanisms, or for the detection of irradiated drugs. Actually, a lot of drugs and excipients can be irradiated in the solid state, and they show no ESR signals in unirradiated samples which are consequently very easy to detect. Moreover, aside from qualitative detection, ESR can be used for dose estimation. Electron Spin Resonance spectroscopy is a suitable technique for distinguishing between irradiated and unirradiated pharmaceuticals, and it can be used for stability analysis of pharmaceuticals. It should be noted that sample preparation is an important step for improved ESR spectral characteristics. Dose estimation from radio-sterilized pharmaceuticals can be obtained by reirradiating the pharmaceuticals with a number of different doses and measuring the ESR signal intensity at each dose interval. Irradiated drugs and excipients may be controlled by ESR by determining the level and stability of drug-derived radicals in the solid state (e.g., tablets, capsules), or in solution. It should be noted that radical species are unstable when dissolved, and the high radical level might be harmful for organisms weakened by, for example, immune deficiency syndrome, cancer, radiotherapy, or chemotherapy.

REFERENCES

[1] W. He, Y. Liu, W.G. Wamer, J.-J. Yin, Electron spin resonance spectroscopy for the study of nanomaterial-mediated generation of reactive oxygen species, J. Food Drug Anal. 22 (2014) 49–63.

[2] W.W. He, Y.T. Zhou, W.-G. Wamer, Mechanisms of the pH dependent generation of hydroxyl radicals and oxygen induced by Ag nanoparticles, Biomaterials. 33 (2012) 7547–7555.

[3] H.B. Ambroz, E.M. Kornacka, B. Marciniec, M. Ogrodowczyk, G. Przybytniak, EPR study of free radicals in some drugs γ-irradiated in the solid state, Radiat. Phys. Chem. 58 (2000) 357–366.

[4] S. Talbi, J. Raffi, S. Aréna, J. Colombani, P. Piccerelle, P. Prinderre, et al., EPR study of gamma induced radicals in amino acid powders, Spectrochim. Acta Part A 60 (2004) 1335–1341.

[5] J. Raffi, S. Gelly, L. Barral, F. Burger, P. Piccerelle, P. Prinderre, et al., Electron paramagnetic resonance of radicals induced in drugs and excipients by radiation or mechanical treatments, Spectrochim. Acta Part A 58 (2002) 1313–1320.

[6] S. Kempe, H. Metz, K. Mäder, Application of electron paramagnetic resonance (EPR) spectroscopy and imaging in drug delivery research – chances and challenges, Eur. J. Pharm. Biopharm. 74 (2010) 55–66.

[7] S. Rana, R. Chawla, R. Kumar, S. Singh, A. Zheleva, Y. Dimitrova, et al., Electron paramagnetic resonance spectroscopy in radiation research: current status and perspectives, J. Pharm. Bioallied Sci. 2 (2010) 80–87.

[8] J. Raffi, S. Gelly, P. Piccerelle, P. Prinderre, A. Chamayou, M. Baron, Electron spin resonance-thermoluminescence studies on irradiated drugs and excipients, Radiat. Phys. Chem. 63 (2002) 705–707.

[9] A. Adhikary, D. Becker, M. Sevilla, Electron spin resonance of radicals in irradiated DNA, in: A. Lund, M. Shiotani, (Eds.), Applications of EPR in Radiation Research, Springer International Publishing (2014), pp. 299–352.

[10] I.A. Bhatti, K. Akram, J.-J. Ahn, J.-H. Kwon, Electron spin resonance analysis of radiation-induced free radicals in shells and membranes of different poultry eggs, Food Anal. Methods 6 (2013) 265–269.

[11] S. Onori, M. Pantaloni, P. Fattibene, E.C. Signoretti, L. Valvo, M. Santucci, ESR identification of irradiated antibiotics: cephalosporins, Appl. Radiat. Isot. 47 (1996) 1569–1572.

[12] A. Bartolotta, M.C. D'Oca, M. Campisi, V. De Caro, G. Giandalia, L.I. Giannola, et al., Effects of gamma-irradiation on trehalose–hydroxyethylcellulose microspheres loaded with vancomycin, Eur. J. Pharm. Biopharm. 59 (2005) 139–146.

[13] J.P. Basly, I. Longy, M. Bernard, ESR identification of radiosterilized pharmaceuticals: latamoxef and ceftriaxone, Int. J. Pharm. 158 (1997) 241–245.

[14] J.P. Basly, I. Basly, M. Bernard, ESR spectroscopy applied to the study of pharmaceuticals radiosterilization: cefoperazone, J. Pharm. Biomed. Anal. 17 (1998) 871–875.

[15] M. Gibella, A.-S. Crucq, B. Tilquin, P. Stocker, G. Lesgards, J. Ra, Electron spin resonance studies of some irradiated pharmaceuticals, Radiat. Phys. Chem. 58 (2000) 69–76.

[16] S. Wilczyński, B. Pilawa, R. Koprowski, Z. Wróbel, M. Ptaszkiewicz, J. Swakoń, et al., Free radicals properties of gamma-irradiated penicillin-derived antibiotics: piperacillin, ampicillin, and crystalline penicillin, Radiat. Environ. Biophys. 53 (2014) 203–210.

[17] A. Skowronska, M. Wojciechowski, P. Ramos, B. Pilawa, D. Kruk, ESR studies of paramagnetic centers in pharmaceutical materials – cefaclor and clarithromycin as an example, Acta Phys. Pol. Ser. A 121 (2012) 514–517.

[18] A.-A. Wassel, Electron spin resonance (ESR) investigation of gamma-irradiated antibiotic azithromycin, J. Adv. Sci. Res. 3 (2012) 49–55.

[19] B.K. Singh, D.V. Parwate, I.B. Dassarma, S.-K. Shukla, Radiation sterilization of fluoroquinolones in solid state: investigation of effect of gamma radiation and electron beam, Appl. Radiat. Isot. 68 (2010) 1627–1635.

[20] O. Cozar, V. Chis, L. David, G. Damian, I. Barbur, ESR investigation of gamma-irradiated aspirin, J. Radioanal. Nucl. Chem. 220 (1997) 241–244.

[21] J.L. Duroux, J.P. Basly, B. Penicaut, M. Bernard, ESR spectroscopy applied to the study of drugs radiosterilization: case of three nitroimidazoles, Appl. Radiat. Isot. 47 (1996) 1565–1568.

[22] G. Damian, EPR investigation of γ-irradiated anti-emetic drugs, Talanta. 60 (2003) 923–927.

[23] A. Özer, S. Turker, S. Çolak, M. Korkmaz, E. Kiliç, M. Özalp, The effects of gamma irradiation on diclofenac sodium, liposome and niosome ingredients for rheumatoid arthritis, Inter. Med. Appl. Sci. 5 (2013) 122–130.

[24] J.-J. Ahn, H.M. Shahbaz, K. Akram, J.-Y. Kwak, J.-H. Kwon, Improved electron spin resonance spectroscopy with different sample treatments to identify irradiated sprout seeds, Food Anal. Methods 7 (2014) 1874–1880.

[25] K.-H. Kim, J.-H. Son, Y.-J. Kang, H.-Y. Park, J.-Y. Kwak, J.-H. Lee, et al., Detection characteristics of gamma-irradiated seeds by using PSL, TL, ESR and GC/MS, J. Food Hyg. Saf. 28 (2013) 130–137.

[26] M. Ukai, H. Kameya, H. Nakamura, Y. Shimoyama, An electron spin resonance study of dry vegetables before and after irradiation, Spectrochim. Acta Part A 69 (2008) 1417–1422.

[27] H. Nakamura, M. Ukai, Y. Shimoyama, An electron spin resonance study of γ-ray irradiated ginseng, Spectrochim. Acta Part A 63 (2006) 883–887.

[28] R. Yamaoki, S. Kimura, M. Ohta, Electron spin resonance spectral analysis of irradiated royal jelly, Food. Chem. 143 (2014) 479–483.

[29] R. Yamaoki, S. Kimura, M. Ohta, ESR characterization of irradiated peony root, a medicinal plant containing crystalline sugars, Radioisotopes. 62 (2013) 631–637.

[30] N. Benbettaïeb, O. Chambin, A. Assifaoui, S. Al-Assaf, T. Karbowiak, F. Debeaufort, Release of coumarin incorporated into chitosan-gelatin irradiated films, Food Hydrocoll. 56 (2016) 266–276.

[31] A. Faucitano, A. Buttafava, L. Montanari, F. Cilurzo, B. Conti, I. Genta, et al., Radiation-induced free radical reactions in polymer/drug systems for controlled release: an EPR investigation, Radiat. Phys. Chem. 67 (2003) 61–72.

[32] B. Marciniec, M. Stawny, K. Olszewski, M. Kozak, M. Naskrent, Analytical study on irradiated methylxanthine derivatives, J. Therm. Anal. Calorim. 111 (2013) 2165–2170.

[33] U. Sayin, EPR analysis of gamma irradiated single crystal cimetidine, J. Mol. Struct. 1031 (2013) 132–137.

[34] P. Ramos, B. Pilawa, E. Stroka, EPR studies of free radicals in thermally sterilized famotidine, Nukleonika 58 (2013) 413–418.

[35] J.P. Basly, I. Basly, M. Bernard, Electron spin resonance identification of irradiated ascorbic acid: dosimetry and influence of powder fineness, Anal. Chim. Acta. 372 (1998) 373–378.

[36] J.-P. Basly, I. Basly, M. Bernard, Electron spin resonance detection of radiosterilization of pharmaceuticals: application to four nitrofurans, Analyst 123 (1998) 1753–1756.

[37] R. Koseoglu, E. Koseoglu, F. Koksal, Electron paramagnetic resonance of some γ-irradiated drugs, Appl. Radiat. and Isot. 58 (2003) 63–68.

[38] F.A. Taiwo, L.H. Patterson, E. Jaroszkiewicz, B. Marciniec, M. Ogrodowczyk, Free radicals in irradiated drugs: an EPR study, Free Radic. Res. 31 (1999) 231–235.

CHAPTER 8

Free Radicals in Nonirradiated and Irradiated Foods Investigated by ESR and 9 GHz ESR Imaging

K. Nakagawa
Hirosaki University, Hirosaki, Japan

Contents

8.1 INTRODUCTION

Free radicals in living organisms are generated during antioxidant activities and biochemical processes. In most cases, there are paramagnetic species in the pigmented regions of plant seed coats. Stable free radicals are also generated by several commonly used industrial processes such as heat treatment or ionizing irradiation, which are well-established procedures used for improving the shelf life of foods. Electron spin resonance (ESR) or electron paramagnetic resonance (EPR) can be used to detect free radicals. The ESR spectrum appears as an asymmetric line shape, or may show multiple overlapping lines, depending on the type of food [1–4].

ESR imaging (ESRI) is an important tool for obtaining information about the spatial distribution of unpaired spins both in vitro, and noninvasively in vivo in samples [1,2,5–7]. The method commonly used to encode spatial information in the data obtained from an ESR spectrometer is the application of a linear gradient to the static magnetic field. The data thus obtained represent the free radical density projection in the direction of the gradient. A two- or three-dimensional image can be obtained by acquisition of multiple projections in multiple directions.

Noninvasive 9 GHz ESRI and continuous wave (CW) ESR were used to investigate the locations of paramagnetic species in various foods. The ESRI can localize the paramagnetic species (transition metal ions, transition metal complexes, and stable organic radicals) in a sample. The distribution of a species in a sample may suggest the origin of the species in nature. The X-band (9 GHz) ESRI has a higher spatial resolution and greater sensitivity than L-band (1 GHz) ESRI. Recently, L-band (~1 GHz) ESRI has been developed to investigate the spatial distribution of spin probes administered to small animals [7]. There have been several reports of X-band (9 GHz) ESRI used to investigate free radicals in naturally occurring samples [1–3]. Thus, noninvasive ESRI and CW ESR can provide quantitative information about specific paramagnetic species. The endogenous paramagnetic species in samples and the application of ESRI to various nonirradiated and irradiated foods are described.

8.2 MATERIALS AND METHODS

8.2.1 Samples

8.2.1.1 Mushroom

Dried (unsliced) and fresh (raw) shiitake mushrooms were purchased from local supermarkets. The mushrooms were used as purchased. Part of a mushroom (0.0045 g) was inserted into an ESR tube for CW ESR [8]. The sliced dried mushroom (0.0184 g) was glued onto an ESR rod for ESRI. In addition, the fresh mushrooms were used as purchased. The cap portion of the dry mushroom is heavier than that of the interior. The fresh shiitake mushrooms contain water.

The water-soluble spin probe 4-hydroxy-2,2,6,6-tetramethylpiperidin-1-oxyl (TEMPOL) was purchased from NacalaiTesque, Inc., Kyoto, Japan. Mushroom samples (pigmented and unpigmented mushrooms) were cut into 10 mm × 10 mm pieces approximately 1.5 mm thick weighing c.0.0405 g. The cut mushroom samples were submerged in 1 mL distilled water for 6 h. The final concentration of TEMPOL was 0.5 mM. The aqueous extract of the pigmented cap was brownish in color. TEMPOL (5 μL of 0.1 mM aqueous solution) containing 5 μL of the water extract was placed in a vial, mixed for 30 s with an agitator, and transferred to an ESR capillary (1.0 mm outer diameter (O.D.)), which was sealed and kept at ambient temperature (approximately 25°C).

8.2.1.2 Sesame Seeds

Various commercial toasted sesame seeds were purchased from a local supermarket. The seeds were used as purchased. The white sesame seeds were irradiated using γ-rays at a dose of 7.3 kGy in 2009. One seed (~0.0029 g) was inserted into an ESR tube for CW ESR measurements [2].

8.2.2 Continuous Wave Electron Spin Resonance Settings

A JEOL RE-3X 9 GHz ESR spectrometer was used for the CW measurements. The system was operated in X-band mode at 9.44 GHz and 100 kHz modulation frequency. All CW ESR spectra were obtained with a single scan. Typical CW ESR settings were as follows: microwave power, 5 mW; time constant, 0.1 s; sweep time, 4 min; magnetic field modulation, 0.3 mT; and sweep width, 10 mT. All measurements were performed at ambient temperature.

8.2.3 Electron Spin Resonance Imaging

The experimental requirements for ESRI are, therefore, the capability of generating linear magnetic field gradients of sufficient strength, and of acquiring multiple ESR spectra in the presence of a gradient of varying direction. Also required is software to perform the necessary processing of the acquired data in order to reconstruct the image.

Reference is made in this chapter to three orthogonal gradient coils and to the imaging direction or orientation. The frame of axes used in this context is a right-handed frame, the Z-axis pointing in the direction of the static magnetic field, B_0. As the majority of ESR magnets have this direction in the horizontal plane, the second convention is that the Y-axis is along the vertical. The frame of axes is illustrated in Fig. 8.1.

8.2.4 Electron Spin Resonance Imaging Settings

Images were acquired at room temperature on a Bruker E500 ELESYS system (Bruker BioSpin GmbH, Karlsruhe, Germany) equipped with a high sensitivity TM resonator (10 mm diameter, Bruker). The system was

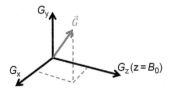

Figure 8.1 Gradient coil frame of axes showing polar angles for an arbitrary direction in this frame. The B_0 indicates static magnetic field.

Table 8.1 ESR imaging conditions for the dry shiitake mushroom and sesame seeds

Conditions	Shiitake mushroom	Sesame seeds
Field of view (mm)	14	6
Pixel size (mm)	0.4	0.15
Gradient strength (mT/cm)	5	12.5
Sweep time (s)	40	90
Total acquisition time (min)	20	50
Modulation amplitude (mT)	0.4	0.5
Microwave power (mw)	5	2

operated in X-band mode at approximately 9.6 GHz and 100 kHz modulation frequency. For imaging, the system was equipped with water-cooled gradients, allowing a magnetic field gradient up to 10 mT/cm along the X- and Y-axes.

For each measurement, the microwave power was selected within the linear section of the power intensity curve. Amplitude modulation values were chosen in such a way that they did not induce any signal distortion, and were always limited to the linewidth value. The conversion time, time constant, field sweep for images, and gradient intensity of images were optimized for each sample, and are given in Table 8.1.

Two-dimensional (2D) images were constructed from a complete set of projections, which were collected as a function of the magnetic field gradient, using the back projection algorithm provided in the Xepr software package from Bruker (Karlsruhe, Germany). Before reconstruction, each projection was deconvolved using fast Fourier transformation with a measured zero-gradient spectrum to improve the image resolution. To reduce noise amplification and avoid possible division by zero at high frequencies, a low-pass filter was used. The deconvolution parameters, including the maximum cut-off frequency and the width of the window in the Fourier space, were set after the shapes of all projections were viewed.

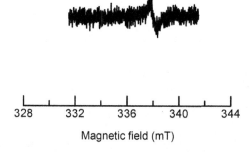

Magnetic field (mT)

Figure 8.2 ESR spectrum of nonirradiated fresh (raw) shiitake mushroom (0.0264 g) is presented. The spectrum was obtained in a single scan.

Spectral deconvolution and filtered back projection were performed using the Xepr software package.

The typical ESRI settings were as follows: microwave power, 2 mW; total acquisition time, 20 min; magnetic field modulation, 0.5 mT; sweep width, 15 mT; field of view, 6 mm; pixel size, 0.15 mm; and gradient strength, 1.6–10 mT/cm. The detailed acquisition parameters for the phantom and mushroom are listed in Table 8.1. All measurements were performed at ambient temperature.

8.3 CONTINUOUS WAVE ELECTRON SPIN RESONANCE AND ELECTRON SPIN RESONANCE IMAGING OF THE MUSHROOM RESULTS

Continuous wave ESR was used to measure the persistent radical species in the mushroom. Fig. 8.2 shows the ESR spectrum from the nonirradiated mushroom, which is a single line of approximately 340 mT. The peak-to-peak line width (ΔH_{pp}) for the raw mushroom was approximately 0.57 mT. Fig. 8.3 shows the ESR spectrum of the dry mushroom, which is a single line. The ΔH_{pp} for the raw mushroom was also approximately 0.57 mT.

Fig. 8.4 shows the sample setting and the resulting ESRI. An image of the cap portion of the mushroom glued onto an ESR rod and a 2D ESRI of (A) are shown in Fig. 8.4B. The spatial resolution of the ESRI of the mushroom was approximately 570 μm based on the linewidth. The 2D image shows that the red portions were regions with high concentrations of the radicals. An image of the stem and cap portions of the mushroom glued onto an ESR rod is shown in Fig. 8.4C, and the 2D ESRI of the samples with this setup is presented in Fig. 8.4D.

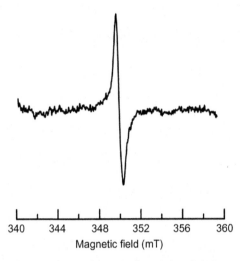

340   344   348   352   356   360
Magnetic field (mT)

Figure 8.3 The CW ESR spectrum of nonirradiated dry shiitake mushroom (approximately 0.0045 g) is presented. The ESR spectrum was acquired at 10 mT sweep width. The spectrum was obtained in a single scan.

The 2D images show an inhomogeneous distribution of the spin density around the cap of the mushroom (Fig. 8.4). The image of the sample mushroom was obtained at ~337.0 mT magnetic field, and was in direct relation to the paramagnetic species. The concentration distributions of paramagnetic species along the cap are shown in Fig. 8.4. No radicals were observed in any other portion of the mushroom, except for the cut stem region.

For the mushroom examined, the ESR linewidths for both fresh and dry mushroom were approximately 0.57 mT. The same linewidth for both samples suggests that the paramagnetic species may be the same. In the case of the fresh mushroom, the microwave may not reach effectively to the species because of the high water content.

The maximum gradient used (10 mT/cm) was sufficient to resolve the mushroom images (Table 8.1). The ESR images of the mushroom shown in Fig. 8.4 show reflected, unpaired spin distribution in a relatively restricted area of the mushroom, the top of the cap. The red color in the image suggests that a high radical concentration is present on the cap. The surface of the mushroom cap has a dark color. It is noteworthy that the intensities of the ESR images were associated with the top of the cap, and were not observed in the interior of the mushroom. Thus, the persistent paramagnetic species were localized to specific regions, such as the cap top and the shortened stem

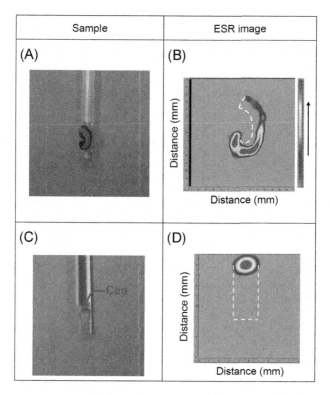

Figure 8.4 ESR experimental set up and ESR images. (A) The cap portion of a mushroom glued onto an ESR rod. (B) The 2D ESR imaging of (A) is presented. The 2D image shows that the *red* (dark gray in print version) portions are intense regions. (C) The cap portion of a mushroom glued onto an ESR rod. (D) The 2D ESRI of (C) is presented. Note that the *white dotted lines* in Figs. 8.4B and D indicate near actual samples.

region. No image was seen other than that of the mushroom cap and the chopped stem region. The weak signal from the chopped stem part implies that paramagnetic species might be related to the food processes used.

Next, we examined the antioxidant activities of the mushroom. Fig. 8.5 shows representative spectra of the TEMPOL solution mixed with the water extract from the pigmented cap region. Water-soluble TEMPOL was used instead of organic solvent-soluble 1,1-diphenyl-2-picrylhydrazyl (DPPH), which does not dissolve in water. Fig. 8.6 shows the reduction in TEMPOL signals as a function of time.

Our primary research interests were the identification of paramagnetic species and antioxidant activities in aqueous solutions from mushrooms. We used TEMPOL for investigating antioxidant activities instead

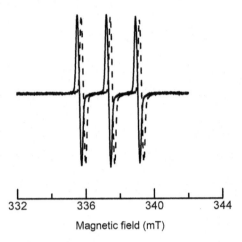

Figure 8.5 Representative ESR spectra of a solution mixture (water extract from the pigmented cap region and 4-hydroxy-2,2,6,6-tetramethylpiperidin-1-oxyl (TEMPOL) solution) in ESR capillaries. The solid and dashed spectra were obtained at 0 and 120 min, respectively.

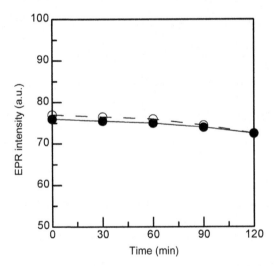

Figure 8.6 Plot of ESR intensities of a mixed solution as a function of time. The filled circle and open circle indicate the extract from the pigmented cap region and the unpigmented region, respectively.

of DPPH, because DPPH does not dissolve in water and we did not use organic solvents to extract compounds from the mushrooms. Although antioxidant reactions may be slow in aqueous solutions at ambient temperature, we prevented organic solvent damage to the mushroom tissues.

Both cap and interior extracted solutions from the mushroom showed slow antioxidant activity. The reduction of the TEMPOL signals (filled circle) by the pigmented solution from the cap, and the unpigmented solution from the mushroom, was associated with antioxidant activity, as shown in Fig. 8.6. The results suggest that other regions of the mushroom also contain similar amounts of water-soluble antioxidant compounds. Although no detectable paramagnetic species or free radicals were found in locations other than the cap region, both extracted solutions showed antioxidant activity, such as the reduction of TEMPOL signals.

For fresh (raw) shiitake mushroom, we observed a very weak signal from the cap (supplemental materials), and almost no signal from the interior of the mushroom. We observed almost no antioxidant activity in the water extract of fresh mushroom, and no reduction of TEMPOL signals in either extract.

The link between in vitro antioxidant activity and the reduction of stable radicals of the extracts of shiitake mushroom have been investigated by several research groups [8–11]. The radical scavenging effects of the organic solvent extracts from the mushroom on DPPH radicals increased with extract concentration. Total phenolic contents were suggested to be the antioxidant compounds in the mushroom extracts. Thus, the total phenolic contents, which have −OH groups, dissolve in water and can be related to antioxidant activity in the cap of the mushroom. However, no previous research has shown stable paramagnetic species in the mushroom; the paramagnetic species may be produced during drying and/or food processing.

The dark color of the mushroom cap may be associated with a biological function. The skin of black sesame seeds [2], black pepper seeds [1], and beans with dark pigmented coat (skin) showed EPR signals in the $g \cong 2.0$ region. The EPR signal was stable and reproducible. In general, compounds that contain stable unpaired electrons are less reactive than reactive oxygen species, because unpaired electrons delocalize throughout the neighboring orbitals. In addition, the dark-colored top (or skin) may offer protection from outside stimulation such as by the environment.

8.4 SESAME SEED RESULTS

8.4.1 Continuous Wave Electron Spin Resonance and Electron Spin Resonance Imaging

Continuous wave ESR was used to measure the persistent radical species of various sesame seeds. Fig. 8.7 shows the ESR spectra of the sesame seeds. The spectrum of back sesame seed shows a single line. The

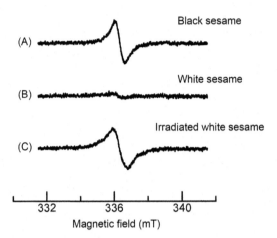

Figure 8.7 Continuous wave ESR spectra of various sesame seeds are presented. (A) Black sesame seed (0.0027 g); (B) white sesame seed (0.0030 g); and (C) white sesame seed (0.0030 g) irradiated with γ-rays. Each spectrum is obtained from a single seed inserted into an ESR tube under the same ESR conditions.

peak-to-peak linewidth (ΔH_{pp}) for the black seed is ~0.58 mT. The white sesame seed showed a very weak signal. The ESR intensity of the white sesame seed was approximately six times weaker than that of the black sesame seed. The irradiated white sesame seed showed a strong ESR signal. The ESR intensities were similar to that of the black seed. We chose to use the irradiated seed because the signal intensity is similar to that of the black seed. The signal was slightly broader than that of the black sesame seed. The ΔH_{pp} value for the irradiated seed was ~0.82 mT.

Fig. 8.8 shows a schematic illustration of the experimental setting for 2D imaging measurements. Two sesame seeds were pasted simultaneously side-by-side on an ESR rod. Fig. 8.9 shows the 2D ESR images of black (top) and white (bottom) sesame seeds. The spatial resolution of ESRI in the black sesame seed was approximately 460 μm. The top images acquired in 2D showed a rather homogeneous distribution of the spin density around the seed. Note that a very high radical concentration was observed at the hilum region of the black sesame seed. In addition, the black coat (skin) part of the seed showed a high radical concentration. Few radicals were observed inside the seed, as is shown in Fig. 8.9. On the other hand, the white sesame seed exhibited the ESR signal and showed few radicals in the seed.

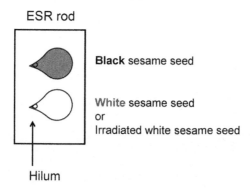

Figure 8.8 Schematic illustration of the experimental setting for 2D imaging. Two sesame seeds are pasted side-by-side onto an ESR rod.

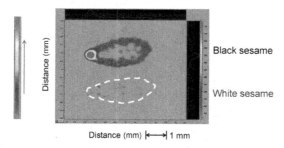

Figure 8.9 Two-dimensional images of the sesame seeds; black sesame seeds (top) and white sesame seeds (bottom). The 2D images of the two seeds are taken simultaneously.

For all sesame seeds examined, the ESR linewidth was ∼0.58 mT. The resolution is approximately 460 μm based on the linewidth. However, the current maximum gradient (12.5 mT/cm) is sufficient to resolve the seed images (Table 8.1). The ESR images of the seeds shown in Fig. 8.9 exhibit reflected unpaired spin distribution in relatively small structures, such as the black colored coat and hilum. The red color in the image suggests that a high concentration of radicals is present in the hilum. Note that the intensities of the ESR images are attributed to the hilum and black sesame seed coat, and were not observed for the region inside the seed. A very low intensity image was observed for the white sesame seed as is shown in Fig. 8.9.

The in vitro antioxidant activities of solvent extracts of the sesame seed coat have been investigated by different research groups [12–14]. The values of total phenolic content of the hull extracts of black and white sesame seed coats were 146.6 ± 4.1 and 29.7 ± 0.9 catechin equivalents/g, respectively [12]. The brown pigment extracted from black sesame seeds using a solvent showed antioxidant activities [14]. The brown pigment contains total phenolic contents, such as lignan compounds. The extracts of the white sesame seed coat containing total phenolic contents also showed antioxidant activities [13]. Thus, the difference between black and white sesame seed coats is the colored pigment containing total phenolic contents. The colored pigment may contain stable paramagnetic species. However, no previous research has shown stable paramagnetic species in the black sesame seed coat. The paramagnetic species may be present orig- inally or produced during harvest and/or food processes.

In a seed, the coat, hilum, and embryo have specific functions or roles. The different concentrations of paramagnetic species found in various portions of a seed may be associated with different functions. The reason for the concentration of paramagnetic species in the hilum of black ses- ame seed has not been clarified.

Fig. 8.10 shows black (top) and white irradiated (bottom) sesame seeds. The radical species of the irradiated sesame seed were distributed throughout the entire area of the seed. The highest intensity of image was observed around the hilum.

The ESR image for the irradiated white seed shows radicals located at the hilum and throughout the seed, as shown in Fig. 8.10. The seed irradiated with γ-rays showed the decomposition of various compounds in the seed. The short-lived free radicals were recombined and became

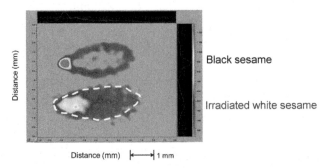

Distance (mm) |←——→| 1 mm

Figure 8.10 Two-dimensional images of the sesame seeds; black sesame seeds (top) and irradiated white sesame seeds (bottom). The 2D images of the two seeds were taken simultaneously.

stable radicals. These results suggest that the broad linewidth of the irradiated seed results from overlapping multiple radicals in the seed. Thus, the ESRI nondestructively showed that the tendency for different parts of the sesame seed to exist as stable radicals varies.

Although ESR is a very sensitive spectroscopic technique, the natural concentration of paramagnetic species which is a key for the nuclear magnetic resonance method is far less than that of protons in biological samples. Thus, satisfactory sample images with sufficiently high resolution may require additional data acquisition. For the ESRI of sesame seeds, we used a data acquisition time of 50 min (Table 8.1). In addition, a higher resolution can be achieved by using a stronger field gradient. A high resolution also necessitates increasing the number of projection angles.

The ESR and ESRI investigations provided quantitative information about the antioxidant activities of aqueous extracts, and revealed the paramagnetic species and concentration distributions in nonirradiated dry shiitake mushroom. Both 9 GHz ESRI and CW ESR are useful for detecting and identifying the location of paramagnetic species in nonirradiated mushrooms and other foodstuffs [2,8].

ACKNOWLEDGMENTS

Part of this research was supported by a Grant-in-Aid for Challenging Exploratory Research (24650247, 15K12499) and for Scientific Research (B) (25282124) from the Japan Society for the Promotion of Science (JSPS) (K.N.), and A-step (AS262Z00876P) from the Japan Science and Technology (JST) (K.N.).

GLOSSARY (NOMENCLATURE LIST)

CW	continuous wave
2D	two-dimensional
DPPH	1,1-diphenyl-2-picrylhydrazyl
EPR	electron paramagnetic resonance
ESR	electron spin resonance
ESRI	electron spin resonance imaging
ΔH_{pp}	peak-to-peak linewidth
TEMPOL	4-hydroxy-2,2,6,6-tetramethylpiperidin-1-oxyl

REFERENCES

[1] K. Nakagawa, B. Epel, Location of radical species in a black pepper seed investigated by CW EPR and 9 GHz EPR imaging, Spectrochim. Acta Part A 131 (2014) 342–346.

[2] K. Nakagawa, H. Hara, Investigation of radical locations in various sesame seeds by CW EPR and 9 GHz EPR imaging, Free Radic. Res. 49 (2015) 1–6.

[3] K. Nakagawa, Effects of low dose X-ray irradiation of eggshells on radical production, Free Radic. Res. 48 (2014) 679–683.

[4] E.D. Seletchi, O.G. Duliu, Comparative study on ESR spectra of carbonates, Rom. J. Phys. 2 (2007) 657–666.

[5] K. Nakagawa, Y. Ohba, B. Epel, H. Hirata, A 9 GHz EPR imager for thin materials: application to surface detection, J. Oleo. Sci. 61 (2012) 451–456.

[6] G.R. Eaton, S.S. Eaton, K. Ohno, EPR Imaging and In Vivo EPR, CRC Press, Boca Raton, FL, 1991.

[7] B. Epel, H.J. Halpern, Electron paramagnetic resonance oxygen imaging in vivo. Electron paramagnetic resonance, R. Soc. Chem. 23 (2013) 180–208.

[8] K. Nakagawa, H. Hara, Paramagnetic species and antioxidant properties in various shiitake mushroom investigated by continuous wave EPR and 9 GHz EPR imaging: Conference Proceedings of 7th Biennial Meeting of Society for Free Radical Research-Asia, Chiang Mai, Thailand (2015) 7–16.

[9] P.P. Levêque, Q. Godechal, B. Gallez, EPR spectroscopy and imaging of free radicals in food, Israel J. Chem. 48 (2008) 19–26.

[10] Y. Ishikawa, K. Morimoto, T. Hamasaki, Falvoglaucin, a metabolite of Eurotiumchevalieri, its antioxidation and synergismwith tocopherol, J. Am. Oil Chem. Soc. 61 (1984) 1864–1868.

[11] L.M. Cheung, P.C.K. Cheung, V.E.C. Ooi, Antioxidant activity and total phenolics of edible mushroom extracts, Food Chem. 81 (2003) 249–255.

[12] F. Shahidi, C.M. Liyana-Pathirana, D.S. Wall, Antioxidant activity of white and black sesame seedsand their hull fractions, Food Chem. 99 (2006) 478–483.

[13] L.-W. Changa, W.-J. Yena, S.C. Huangb, P.-D. Duh, Antioxidant activity of sesame coat, Food Chem. 78 (2002) 347–354.

[14] J. Xu, S. Chen, Q. Hu, Antioxidant activity of brown pigment and extracts fromblack sesame seed (*Sesamum indicum* L.), Food Chem. 91 (2005) 79–83.

INDEX

Printed in the United States
By Bookmasters